The New Cyberwar

ALSO BY DENNIS F. POINDEXTER

*The Chinese Information War:
Espionage, Cyberwar, Communications Control
and Related Threats to United States Interests*
(McFarland 2013)

The New Cyberwar

*Technology and the
Redefinition of Warfare*

DENNIS F. POINDEXTER

McFarland & Company, Inc., Publishers
Jefferson, North Carolina

LIBRARY OF CONGRESS CATALOGUING-IN-PUBLICATION DATA

Poindexter, Dennis F., 1945–
 The new cyberwar : technology and the redefinition of warfare / Dennis F. Poindexter.
   p.   cm.
 Includes bibliographical references and index.

 ISBN 978-0-7864-9843-7 (softcover : acid free paper) ∞
 ISBN 978-1-4766-2061-9 (ebook)

 1. Cyberspace operations (Military science)   2. Tactics.
 3. National security.   I. Title.
 U167.5.C92P65 2015
 355.4—dc23   2015027210

BRITISH LIBRARY CATALOGUING DATA ARE AVAILABLE

© 2015 Dennis F. Poindexter. All rights reserved

No part of this book may be reproduced or transmitted in any form or by any means, electronic or mechanical, including photocopying or recording, or by any information storage and retrieval system, without permission in writing from the publisher.

Front cover image © 2015 Photodisc

Printed in the United States of America

McFarland & Company, Inc., Publishers
 Box 611, Jefferson, North Carolina 28640
 www.mcfarlandpub.com

TABLE OF CONTENTS

Acknowledgments vi
Introduction 1

1. A New Outcome for War 9
2. A Cloudy Day in Kiev 22
3. The Last Great War 40
4. The War Peeks Out 50
5. Cyberwar 67
6. Economic War 92
7. Economic Espionage 96
8. The Evolution of Command and Control Warfare 104
9. Psychological Warfare 110
10. Follow the Money 117
11. A Military Left Out 130
12. The Integration of Business and War 141
13. The Combatants 164
14. Fighting a Modern War 175
15. The New World War 188

Chapter Notes 193
Bibliography 206
Index 219

Acknowledgments

I owe my ability to write to my wife, Virginia, who agreed to work so I could devote myself to it. She is my inspiration and love each and every day.

I want to thank my brother Gary for graciously agreeing to edit this book. He introduced me to information technology in 1974, and has been a leader in his field for many years. His insight, especially in his experiences in Russia, was very beneficial. His patience was above and beyond the call of duty.

Introduction

When the Russians took Crimea without so much as a peep from Europe or the United States, many thought it was odd. How can one country take another part of a neighbor's territory, and the two of them not be at war? We have to think that over a little.

In the days of the modern wars, the battle lines would be drawn and armed forces of both governments would collide on several fronts. Nothing of the sort happened in Crimea, however, because this is a new kind of war where nobody fights with an army, but both sides fight with the weapons of war. They don't even call it war. We are transitioning from kinetic wars to information wars, but we are not yet sure of what that means. Before long, we will lose a war we didn't even know we were fighting. That war is not about conflict in Crimea, or the Middle East, or Africa, where most of the fighting in the world is going on. It's bigger than that.

Every country fights wars without the consent of its citizens, and often without their participation, but these wars are different, because they are fought without citizens' knowledge. The old wars were public because there was no way to hide what was being done. Soldiers donned uniforms; horses or tanks moved out in columns; airstrikes left visible marks on towns and villages. The new wars offer little to see, and much of that is deception.

What we saw in Crimea, and in the eastern part of the Ukraine, was an almost transparent attempt to use cyberwar to win a major conflict. It isn't supposed to be like that, and the Russians partly failed to achieve their objectives in that respect. Done properly, the key performance parameter is stealth, backed up with credibility. They certainly missed the former, making the latter difficult to sustain. If, in spite of their denials, we still know they are the ones behind the action, then it is not successful cyberwar. Nicholas Burns, former U.S. ambassador to NATO, says the Russians are engaged in a full-scale invasion of eastern Ukraine, dropping even the pretext of not being involved there.[1] Cyberwars are supposed to be fought in the black. This one was a light shade of grey.

The world is full of wars to confuse us into thinking that nothing has changed, but these are not wars like the modern version of cyberwar. Governments fight wars, not terrorists. The wars that terrorists fight are not very satisfying to the terrorists, or to the people they fight over, and only relatively recently have the participants started to use the weapons of modern war. The Islamic State in Iraq and the Levant (ISIL) uses the same tactics that the Russians used in Crimea, but the Russians have been doing it longer and know how it is supposed to be done. While they are far from subtle, they make ISIL look crude and barbaric.

Cyberwars are about the will of the people involved in the fight, and those on the outside looking in. This is not my idea. It comes from Carl von Clausewitz, who described the purpose of war as "the compulsory submission of the enemy to our will." Without that, wars are longer and less satisfying.

Terrorists occasionally ignore the real purpose of war. When they do, they have to deal with the bell-shaped curve of perception. A few people will always favor what they are doing, even the gruesome beheadings that have established a kind of style for ISIL. A few others will vehemently oppose it. Those who favor or oppose are easy to deal with. The terrorists reinforce the ones who favor their moves and stifle the ones who don't, often by killing them. The majority, in the middle of the curve, represent a myriad of opportunities.

Managing the will of that large group requires a sustained and pointed effort to control media in the area of conflict. China is the best in the world at doing that, but Russia and a few of its allies are getting better. No country can control all aspects of the media, but to a certain extent they can manage it. That is what a good bit of this kind of cyberwar is about.

If someone had said "cyberwar" in 1995, we would have known what it meant. The military arms of most governments had doctrines that described it well enough for any student of war to understand. It was part of information warfare, and very narrowly defined. It was an adjunct to kinetic war, which helped to deny or undermine the military capability of an adversary, so those airstrikes and tank battles could be successful.

It is more confusing today. The targets and objectives have changed so radically that wars are not even being fought with soldiers, the information war is over, and the cyberwar is about to begin. The vision for a new kind of war is here, but technology has not caught up to it yet. To a great extent, this is war not fought by the militaries of the world. It is a war we are losing because our adversaries understand it better than we do, and they are capable of waging it.

Because it is in transition, parts of cyberwar are not well defined, whereas information war had doctrines that described it clearly. In military doctrine, information war had these components:

1. Economic warfare: the manipulation of information exchanged in trade (either denial or exploitation) as an instrument of state policy;
2. Command and control warfare (C2): the use of conventional or enhanced methods to attack the enemy's ability to issue commands and exchange them with field units;
3. Electronic warfare: the use of specialized methods to enhance, degrade or intercept radio, radar, or cryptography of the enemy;
4. Intelligence-based warfare: the integration of sensors, emitters, and processors into a system that combines reconnaissance, surveillance, target acquisition, and battlefield damage assessment;
5. Psychological warfare: the use of information to affect the perceptions, intentions, and orientations of others;
6. Cyberwar: the use of information systems against the virtual personas of individuals or groups;
7. Hackerwar: the use of techniques like specialized software to destroy, degrade, exploit, or compromise both military and civilian information systems.

This is total war, and it reflects how war has changed. We no longer have just the elements of force involved in beating an opponent, forcing them to accept our will. Cyberwar, in its more current practice, has the same elements as information war. Use of the term is expanding due to the extension of technology (and its influence on protagonists). As communications capabilities have grown, the ability to apply cyber techniques to other traditional areas of information war has allowed the term *cyberwar* to be applied to more areas of information war. They probably don't have to be, but they are.

I use the example of a billboard, showing a map of Crimea covered in a swastika, which is a psychological warfare technique in information war. The billboard itself may have been drawn up on a computer, electronically sent to a graphics shop, and printed on paper. The billboard is not digitized or electronic in its final form, and it is not read by a computer user. The same image is sent to newspapers to be printed and sent to homes for viewing. It is on websites of several news distribution systems. It is still psychological warfare, but where it is delivered and viewed electronically, it can be thought of as part of cyberwar. The message is the same one on the billboard, but the delivery medium is different.

So, you can ask yourself if the distribution of a newspaper image is cyberwar or not. It doesn't fit the definition of cyberwar in information warfare. But when the image is electronically captured in a digital camera, sent to the *Financial Times*, and electronically published in the paper, it becomes part of something else. It allows us to quickly show how the Russians were able to influence people in that part of the world. Other governments show those images as evidence of what the Russians are up to. Both sides are trying to win the will of the people, and those images are part of it.

Cell-phone videos captured the moment when a commercial airliner was falling out of the sky, with clothing, debris, and papers all tumbling down like snow. The person who took that video was not thinking about cyberwar, and was not part of a government campaign to spread images that might affect how the nationalists were perceived. Somebody gave, or sold, those images to CNN to broadcast, and before the day was out, they were on TV in every part of the world. Combined with cell-phone pictures of a mobile missile platform heading off in the direction of Russia, and Twitter recordings of nationalists who realized their mistake, we could form our own opinions about where the missile came from, where the launcher came from, and who did the shooting. The Russians told a different (and far less credible) story.

We think of that as the influence of a free press on a government's activity, but we could also think of it as part of the war between governments, with the press occasionally being the vehicle of delivery. The Russians, Chinese and others believe that it is important to control their press (the Russians certainly did their best to control it in the Ukraine). Other governments rely on a free press spreading information quickly, in the belief that the truth will win out. That is the real essence of this war. A few countries believe that anyone with influence should be controlled, in order to serve the best interests of their people. But they aren't content to limit that control to their own borders. They want to define what is in the best interests of not only their own people, but ours as well.

Whether you call it information war or cyberwar defines where you are in this transition. I have always called this phenomenon information war and was reluctant to change. My best friends thought I was behind the times, so I had to look more closely at how information war was changing. It wasn't. Instead, cyberwar is growing, absorbing most of what was originally defined as information war. It is partly because technology is growing at such a rapid pace, but also because governments are learning to use that technology to their benefit.

There are three technological trends that drive the transition to cyberwar: increased use of the Internet, the Balkanization of the Internet, and the addition of many things, like connected appliances, self-guided cars, and video feeds from everywhere. This is the Internet of Things. In places like the Ukraine, only about 40 percent of the people have an Internet connection, but with the current rate of the expansion, most of them will have one in another 10 years. A country like Russia would have a hard time fighting a cyberwar with people who have no computers, so they have to use other techniques, like billboards and print media. The time will come when those are no longer necessary. That could be good or bad for the Ukraine, depending upon how governments, including their own, can influence the will of the people who live there.

In Carl von Clausewitz's time, that was done by force, or *violence*, as he called it. In 1832, he had not seen the methods of persuasion that would come to pass. These days, winning the will of the people does not require violence, unless the protagonists are impatient. We can look around on any day and see people who are. What we don't see is cyberwar.

Cyberwar, by my definition, is that part of war that incorporates both the control and the use of information to influence the will of the people, not limited to parties of the dispute. It has elements of information warfare (its roots: economic, psychological, command and control, electronic, intelligence-based, and what used to be called cyberwar), but has now become the name of the whole way of waging war. To say there is some confusion about what it is would be an understatement. There are over 35 definitions of *cyber war* or *cyberwar*, but it can be defined by what we see as its effects.

The Russians did quite a few things right in Crimea, and that little skirmish is far from over. Some 200 people still lose their lives there every week. The Russians won the will of the Russian-speaking people in the region, muting dissent of those who were against their coming. And they are not done with Eastern Europe. One of the characteristics of modern cyberwar is a clear absence of borders; media skip over them like they weren't even there.

Cyberwar is the main reason that conventional wars are not very satisfying to the winner. It is likely to start long before hostilities open, and to continue long after the fighting has stopped. It is an incremental building of collective ideas; it takes time. It is about the will of the people, and not always about territory or political ideology. If it is done correctly, there is little need for killing. People can be convinced that cooperation is beneficial to their common needs.

The press talks about this war in strange ways, like it was a conscious effort on the part of another country to slip something by the rest of us through propaganda. The central theme in the Ukraine was the neo–Nazis taking back parts of the Russian-speaking world, conjuring up images of marauding Nazi hordes coming at these poor people, as they did in World War II. Many people in that part of the world still remember the Nazi threat. That effectively muted, and frustrated, Germany, which is the strongest European country in NATO. People in the neighborhood thought they might be better served to stick with the Russians who were already there.

The press acted as if it was something the Russians decided to do in the last few months, or recently invented to help their cause in Ukraine. Nothing could be further from the truth. Propaganda, a piece of psychological warfare, is only a small part of a cyberwar. Russia and China have been good at it for a long time. The Internet makes them better.

The Russians are following the strategy of China, and they are not the only ones doing so. The Chinese are not officially at war with anyone. They deny that they use war in any way, and proof that would require them to admit it is hard to come by. It is like the criminal who rarely gets caught. He leaves few traces of his crimes—no fingerprints or DNA, no disorder that would be noticed, no visual images of his face, and no evidence linking him to the crime. The police and legal teams have to prove (sometimes circumstantially) that the person on trial is the culprit. His "clean" past and claims of innocence make it easy for the jury to believe he didn't do it and acquit him, even if he is guilty. If he is convicted, it is easy for the judge to be lenient with the prisoner. In either case, he will be back out on the street very soon, and he will be more careful next time.

Mao Zedong recognized this early on in the Chinese revolution when he said:

> For the revolutionary war is a war of the masses; it can be waged only by mobilizing the masses and relying on them.... If we only mobilize the people to carry on the war and do nothing else, can we succeed in defeating the enemy? Of course not. If we want to win, we must do a great deal more. We must lead the peasants' struggle for land and distribute the land to them, heighten their labour enthusiasm and increase agricultural production, safeguard the interests of the workers, establish co-operatives, develop trade with outside areas, and solve the problems facing the masses—food, shelter and clothing, fuel, rice, cooking oil and salt, sickness and hygiene, and marriage. In [short] ... all the practical problems in the masses' everyday life should claim our attention.[2]

Sometimes it is not about persuading an adversary that their beliefs are wrong, but rather using their beliefs to restrain their resistance, or even to cooperate as allies. At other times, it is like sucking the life out of them, making them focus on basic survival. That keeps them busy until the war is over.

I have written this book because what I see coming in cyberwar scares me. I have worked in cybersecurity since the early part of 1977, when it was relatively safe to be out in networks and war was easy to understand. That was the Viet Nam era, when napalm fried a path through the jungle and bullets flew in record numbers. We saw it clearly. That has changed. The Internet has become a way to get to my personal information, my work, and the core of my country, and wars are not wars anymore. They have become invisible.

More important, I have lived my life in what the world calls a free society and would not like to change to what I see being practiced in the name of freedom. Domestic politics has taken on an appearance of cyberwar, and that is disturbing. Political leaders should know better than to tolerate techniques of war used on their own people.

Ask any dictator or authoritarian ruler, and they will say their citizens are free, even though the evidence suggests otherwise. This book is about the evidence, circumstantial though it might be, that some of them are at war with us, and we are losing. It is a war below our threshold for war, because that is what our enemy intends. The first principle of cyberwar is to prevent the perception that we are at war.

1

A New Outcome for War

One of the characteristics of modern war is its inability to lead to a lasting conclusion that favors the winner. A win is an illusion, partly created by leaders who understand war, and partly by militaries left out of it. We still kill people who are the youngest and strongest of our nations, but we aren't at war when we do it. These days, war is not a battle of artillery as much as it is a battle of wills. We are hardly ever at war in the way most people understand.

The knock-down, drag-out fights of the last 200 years were often structured to result in the greatest casualties to the participants. Generations of families were affected, as wars killed the finest young men and women a country had. The Last Great War was the Allies vs. Iraq, a long and winding affair that Iraq lost. Today, it might be difficult to tell exactly who won.

What we see is a new war, an attempt to control the will of populations. A war is not won until that is accomplished. Our enemies understand this concept and attempt to control the will of their own people before trying to influence others. It is a struggle for some countries trying to do it well because the Free World is not suited to modern war. Our enemies use this to their advantage whenever they can.

The Chinese developed and refined information war into something that works fairly well. The book *The Chinese Information War: Espionage, cyberwar, Communications Control and Related Threats to United States Interests* describes how they manage to stay ahead of the rest of the world, and win. The Chinese are never at war with anyone. While we continue to fight kinetic wars that will not be won, the Chinese have not stopped, and they are teaching others how to do the same thing. In Syria, Crimea, and Afghanistan, war is changing because of it. This is cyberwar bleeding its participants. It is subtle, folded in layers of words and pictures, incremental, and very successful.

The Russians adopted the Chinese strategy to take back Crimea after a bloody and public repudiation of their policies in the Ukraine. They were not very good at it. The Ukrainian leader, Viktor Yanukovych, was on the verge of turning his country toward Russia when a popular uprising threw him out. The Ukrainians' ability to remove a leader made the Russians pause and reassess, but not for long. Better than most, the Russians understood what was happening.

For a time, it appeared they had lost out to the will of their neighbors in the Ukraine. However, they had not forgotten their objective, or how to achieve it. They understood how to win, even when they appeared to have lost. What we didn't recognize was that this war in Crimea, and probably the eastern part of the Ukraine, was won before it started.

The Russians denied they had inserted troops on the Crimean peninsula. They didn't fire a shot at anyone. Their first targets were military bases, the oil and gas infrastructure, television and radio stations, and dissidents in their own country.[1] In an amazing coincidence, 14 websites and popular online news broadcasters, who were said to be less responsive to self-censorship, were attacked by a denial of service coming from a Russian hacker group that admitted responsibility.[2] The Russian government closed down websites of groups that might object to what they were doing. They put new leadership in place, held popular votes in record time, and declared victory before the European Union or the United States did much of anything to stop them. It was neither spontaneous nor careless. At least in Crimea, it was a clear triumph for the Russians. They might have done better had they stopped there.

A *Financial Times* article showed a billboard in Crimea, with that part of the Ukraine covered by a swastika. The campaign message, to those who remember World War II, was obvious: the government in Kiev means a return of the Nazis. Crimean radio and television stations that favored the takeover were allowed to operate, and others shut down.[3] There is evidence to suggest the Russians monitored the communications of Ukraine's leadership long before the war started.[4] News outlets in the Ukraine attribute the problems there to "neo–Nazis," a clear image of the Nazi–Soviet Union conflict in the Ukraine. The theme was used in Germany to keep criticism of the takeover to a minimum. The citizens of these countries have parents who were alive when the Nazis came through. They haven't forgotten what happened.

The Russian forces should have been careful to avoid having their communications monitored by others, a lesson learned in Chechnya, where rebels monitored cell phones, radios, TV transmitters, and the Internet.[5] They were not as good as they would have liked, with some soldiers being careless. For instance, a Russian soldier posted photos on his Facebook page. He forgot that cell-phone photos have geo-location data attached to them. His pictures clearly showed him moving back and forth between Russia and the Ukraine.

1. A New Outcome for War

The Russian leaders knew better, and were probably trying to find him minutes after the story ran on CNN. The nationalists were worse. They tweet, and use unsecured telephones that are monitored by others. The Russians and their supporters have to do better if they expect to be successful.

This was a well-planned action, however, in spite of the mistakes. The will of the people, essential to winning wars, cannot be won with military strength alone. This war was won long before the kinetic fighting started. This was an information war–turned–cyberwar, played by new rules, and it is only getting started. We see photos of women holding up signs that show New Russia, comprising a third of the Ukraine on today's map, maybe more tomorrow.

According to the Russian view of things, there was no war going on in Crimea or Ukraine. In the first few weeks, no shots were fired; before August 2014, no tanks were rolling through the streets, as they were in Kiev before Viktor Yanukovych fled the country like a thief in the night. Molotov cocktails flew there, like sparklers on the Fourth of July. When the cover story finally broke down, amid the debris of a civilian airliner, all hell broke loose.

We have to think about this for a bit. We should wonder why a country's leaders would deny involvement in moving troops into an area when they are obviously doing it. That's because they didn't practice their trade very well, and in the end they had to give up the pretense. Too many things happened to undo plausible deniability for their actions. These are critical errors on this type of covert battleground. To make matters worse, their explanations for events were incredible, in a literal meaning of the term.

After Malaysian Flight 17 went down, the Russians took us to a "command center" that looked like it had never been used. The voices of people echoed in the huge room. Very few personnel were on duty, and there was not a thing that would indicate a military group worked there. They offered a theory that another aircraft, an SU-25 flown by the Ukraine, might have shot down the commercial airliner, mistaking it for Vladimir Putin's aircraft. We asked what Putin was doing there, but nobody answered. The Russians postulated that the plane was filled with people who were already dead and shot down to discredit the nationalist forces. These are literally incredible stories, told at a time when credibility counts.

These calculated troop movements were not likely a spontaneous uprising of local militia; yet both Russian senior leaders and their diplomatic representatives denied Russian involvement. The press, or at least the press reporting from the Ukraine, seemed to agree that there was a question about who the protagonists really were. That important concession gave them time to construct better stories, but they didn't use that time very well. Ukraine could erupt in civil war and the Russians would claim its countrymen were at risk. Is this an insult to our intelligence?

Credibility is a key ingredient in any strategy to sway a population's will. People have to believe in the information source, or believe the argument the source makes, to establish belief. When Russia's leader says he is not sending troops into the Ukraine, many people will believe him because of who he is. He can argue, as he was forced to in late August, that those troops of his wandered across the border by mistake. Those kinds of mistakes do happen. But too many other things happened that were incredible long before those troops drifted into the country. When large objects like tanks and missile launchers are coming down the roads from Russia, they are hard to deny. Those kinds of events made the Russian arguments less reliable. Lost is the truth that quite a few armed forces were already in Crimea.

According to a 2010 military agreement, the Russians already had 25,000 troops, 388 ships, and 161 aircraft in the area. They weren't trying to hide them from anybody.[6] What they did was move troops, tanks and aircraft into areas of western Russia, where satellites could see them clearly waiting. NATO released photos showing their presence. Russia said they were old photos and the troops had since come home. NATO then released new "before and after" photos showing the troops had not been there long.[7] We spend half our time denying blatantly false accusations that are easy to refute, but we aren't getting ahead of the game the Russians play. We fail to understand that truth is neither necessary nor sufficient in cyberwar.

The troops that showed up in Crimea have been called "special forces" by our Secretary of State, a term that implies they are not regular army. This strategy allows the Kremlin to say, "We don't have troops in the Ukraine," because special operations forces are not regular army. Use of the term "special operations forces" instead of regular army provides plausible deniability for those deploying the troops. This logic may not seem sound, but every country plays the game the same way. The term "special operations forces" usually applies to small groups of individuals who have to operate like the *Mission Impossible* crews—the Secretary will disavow them if they are caught. They may speak Russian, but so do a third of people in the Ukraine. They may have Russian uniforms, but they got those when they worked for the army; they are discharged, or on leave, from their regular army units. They may have Russian weapons, but these were obviously stolen from an armory somewhere. It is possible to build a believable story, but impossible to sustain it. All along the path of belief is a string of denials. Deniability is essential to any covert operation, but so is credibility of those denials.

In March and August 2014, several Russian forces were arrested in Ukraine. One soldier was in Chonhar at a temporary checkpoint, wearing a special forces black uniform with no insignia. He carried an AKS-74 assault rifle, five magazines, and several ID cards, all in different names.[8] One of the cards indicated his membership in the Military Intelligence of the Armed

Forces of the Russian Federation, also known as Spetsnaz, special operations forces trained in intelligence gathering and special military operations. Why he carried his ID card is a reasonable question. This would be like a spy carrying his government identification badge with him on a trip to another country.

A Russian source showed the *Financial Times* photographs of a memorial service in Moscow for twelve special forces operations troops killed in fighting in the Ukraine.[9] These people are noted to be separate from the Russian mercenaries, former Russian servicemen, and Chechens who fought with the uprising. We would like to know more about those fighters and where they are, but there is nothing more to be found.

Russia calls fighters who favor Russian leadership *nationalists*, though exactly what that means is not clear, nor why it would be significant if they were. Nationalists of what, we could ask? They carry weapons that are not available in the marketplace, and that were used by the Russian forces in Crimea.[10] They seize television and radio stations, as well as military installations, and set up checkpoints on roads. Russia's denial is a little thin on logic, given the number of troops involved and the presence of a free press hounding their movements. But, in the face of evidence to the contrary, these facts may not be enough. Those who want to believe don't need much help.

Bodies are buried by both sides by the nationalists who manned a checkpoint where they were killed, and by the relatives of Kiev supporters who have been tortured, their bodies floating in a river, much like those of Saddam Hussein's opposition in Iraq. The leader of the Committee of Soldier's Mothers says mothers of soldiers are reporting that their sons have disappeared because "they are bringing back dead bodies and burying them in secret." This mimics what happened in Afghanistan and Chechnya.[11] In addition, armed nationalists take a building and hold the police chief.[12]

The groups that occupied buildings in the city of Donetsk, in eastern Ukraine, created trouble that wasn't spontaneous, and the White House said they may have been paid for their efforts.[13] A woman, identified in a British newspaper, has appeared in five different locations as a soldier's mother, an aggrieved housewife who had to move to Russia to get away, and a citizen of Kiev who wanted Russia to help her country.[14] She claimed the Russian-speaking people of Ukraine were being discriminated against, and Russian television was on the scene to make sure their view of events was properly structured and distributed.

Kiev said it arrested Russian intelligence operatives working there.[15] Russia denied, to anyone who asked, that these additional troops were theirs, or that they had any agents who were captured.

In April 2014, the Russians agreed to a diplomatic solution requiring the Donetsk thugs to leave their government buildings and checkpoints; they refused. Russia claimed they were Russian nationalists, not controllable by government leaders. They could act surprised at their good fortune, as incredible as it may seem.

A miscalculation occurred in June 2014, when someone left the keys in several Russian T-64 tanks, and the nationalists ran off with some. The United States said these were photographed by NATO being loaded onto trucks in Russia, and then carried across to where they were discovered.[16] These were too big to hide and required some skill to maneuver on narrow roads. Trained operators sprang up and took over. Unfortunately, also nearby were some deadly versions of hand-held anti-aircraft missiles made for shooting down anything that flies relatively slow. They shot down two military planes, killing at least 50 Ukrainian soldiers. The Russians are far from subtle.

A Russian court also had closed-door hearings on charges filed against the Ukrainian Interior Minister, Arsen Avakov, for murder and "the use of banned ways and methods of warfare."[17] This vague term comes from the International Committee of the Red Cross and treaties on the use of war materials that cause unnecessary suffering[18] (a confusing concept, given the wars that we see all around us). The court issued warrants for Ihor Kolomoyski and Avakov; both were tried *in absentia*. Few understand this as cyberwar, but it allows the Russian press to call these people "convicted criminals" and to have them placed on Interpol's wanted list. That information makes them criminals in the eyes of many people, and limits their travel to certain parts of the world. A little sliver of information, generated by a court, will follow them forever.

Ihor Kolomoyski got the Russians' attention when the Ukrainian oligarch began buying supplies for his country's armed forces in the east, spending an estimated $10M of his own money. Kolomoyski's fortune came from banking, and he owns PrivatBank, the largest bank in the Ukraine. Although he lives in Switzerland, he came back to the Ukraine when he was named governor of eastern Ukraine. He doesn't like Putin and they have exchanged pleasantries. He called Putin a "schizophrenic of short stature" (Putin is only 5'7") and "insane." Putin called him "a unique imposter" and probably a few other things outside earshot of the press.

Then, to make the point clear, Russia put the Moscow subsidiary of Kolomoyski's bank under temporary administration, saying it was having liquidity problems.[19] At the end of June 2014, a group of hackers called the Green Dragon Crew claimed it had brought down the website of PrivatBank because of its support for the Ukraine, and the site was "currently unavailable."[20] Ukraine's four main banks have been barred from operating in Crimea, but Russian banks have not kept pace with demand, creating a nervous condition among citizens in the new part of Russia.[21]

Avakov, who appears on nationalist posters with half his face looking like the Two-Face character in Batman comics, is not a shy person. He actively engages on Facebook and is widely read in his own country, usually baiting the Russians in some way. He was first to announce the Russian invasion of Crimea, and claimed that this incursion was the first invasion ever announced on a Facebook page.[22] Being a Russian criminal is not going to bother him very much.

However slanted we may view the election results in Crimea, they clearly favored the Russians. It gave them credibility, deserved or not, which they may hope to duplicate in the Ukraine. The formula is to undermine the government with events they can't control. They have Russian-speaking demonstrators who make demands for Russia's help. They have troops ready on the border to "protect Russian nationalists," who may or may not hold Russian passports. They send "humanitarian aid" in white trucks just like the UN and Red Cross. None of this matters, because they make sure their press cameras capture the happy faces of these liberated people. It is just like making a formulaic movie, only the characters are real. We have seen it before.

This is the third installment for the Russians, so historians might view their moves with less credulity. The first was in Georgia in 2008, when the Russians inserted themselves between two protectorates, Abkhazia and South Ossetia, and the Georgian government. There was real fighting in this intervention, with Russian strategic bombers and fighters shooting up the landscape, launches of a few long-range missiles, and Russian forces attacking Georgian army positions.[23] There was no question about Russian troops being there.

Although they looked like bullies, the Russians now control both parts of Georgia. Most of the world called this a war, though it didn't last very long, and Georgia was not big enough to make much of a lasting impact on the Russian forces. The Russians, however, didn't call it a war. We could do very little about it at the time, and have done next to nothing since, so we didn't call it a war either. We quickly forgot about the whole thing.

The second act took place in little Moldova, one of the poorest countries in the world, where 2,500 armed Russian troops lived as peacekeepers, protecting arms stockpiles, Russian citizens, and friends.[24] Mr. Putin criticized the country for coming to the European Union for some of its energy needs. A stretch of border with the western part of the Ukraine has a large population of Russian-speaking people. It is not Moldova, though it is an area that is ethnically Moldavian—a place where "the hammer and sickles are ubiquitous and the feared secret police are still called the KGB."[25]

The Russians were going to open a consulate in Transnistria, a little sliver of a country between Moldova and the Ukraine, and the Moldavian government said no. If this sounds familiar, it should. It is the Crimean model. By treaty, the Russians have 7,000 troops in Tajikistan, a small country that wraps around the northeast corner of Afghanistan, and less than 5,000 in six other countries.[26] They can easily duplicate their Crimean strategy in any of these locations. The Russians are positioned to use cyberwar to take over these seven countries without the use of kinetic war machines, or even the firing of a single shot.

Vladimir Putin got his practical experience in a real war with Afghanistan, which ended in 1989, and more recently in Chechnya, which is almost over. In 1998, Boris Yeltsin made Putin the chief of the Federal Security Service, and then Prime Minister two years later. In March, the Chechens kidnapped the Russian envoy, General Gennady Nikolaevich Shpigun, holding him for ransom, which set the stage for relations between the Russians and the Chechens. For Putin, it was not a good way to begin, and for Shpigun it was worse (two years later, they found him dead).

Chechnya's leader, Aslan Maskhadov, was far from a radical Islamist. He was in fact a former Soviet Army colonel, returned home to lead the army to a victory over the home country during the first war. He tried to talk to Putin, but Putin told him Boris Yeltsin said he was not to speak with him. Maskhadov was clear in blaming Putin for much of what was happening in his country:

Maskhadov went on to say "already at the beginning of this war it was clear that it was impossible to confine it within the limits of Chechnya. The same sort of punitive operations that were launched in Chechnya also began in Ingushetia, Dagestan, North Ossetia, and Kabardino-Balkaria. It was the Federal Security Service that inflicted the war on those republics," not Osama bin Laden or Al-Qaeda. Maskhadov said he is certain that "bin Laden couldn't even find Chechnya on a map."[27]

What followed was open war, which ultimately cost the Russians 5,200 men in a country less than 10 percent the size of Afghanistan, where they lost 13,000.[28] They killed tens of thousands of civilians. The Russians were condemned for human rights violations and the Obama Administration sanctioned a few of them, long before the current crop of sanctions started.[29] From Putin's perspective, not having a war must be preferable to having one. Better not to be at war with anyone.

Is Russia insulting our intelligence?

Yes, in a way. We can argue that our intelligence agencies should tell us that this is a Russian invasion and we should act accordingly. They may have, but Western governments have not acted like it.

When the latest moves began in the Ukraine, the United States asked the European Union for sanctions, but there seemed to be no urgent need to enact sanctions. The U.S. Senate went on vacation before voting on an aid package for the Ukraine. The British have complained that sanctions might even cause more harm to them than the Russians. Europe is concerned about follow-up sanctions that may affect their own countries as much as Russia. German industrialists like BASF, Siemens AG, Volkswagen, and Adidas, and Deutsche Bank balk at more sanctions.[30]

McDonald's took the brunt of the reaction, when the Russians claimed there was high bacteria content in its food. The Russian Rospotrebnadzor, which oversees food standards, said it found E. coli bacteria in McDonald's Caesar wraps and vegetable salads and, by the way, had underestimated its calorie counts in burgers and shakes. McDonald's, which has operated in Russia for 25 years, was perplexed at why it was singled out, but knew what this was about, and it wasn't calorie counts. Ukrainian milk suffers the same fate as Georgian wine and Polish pork. When a country crosses the Russians, there is a price to be paid.[31] They block the sale of Ukrainian vegetables in Russia because they are "improperly labeled." The price is selective enforcement of laws against countries who don't agree with a Russian position or who attempt to impose their will on the Russian government. The Chinese do the same thing.

For weeks, nobody raised a hand to stop this force from entering Ukraine; no hostile shots were fired from either side. Until an announcement by the United States, a month later, that it was sending troops to Poland, Latvia, Lithuania and Estonia, no troop movements were seen from NATO, the obvious choice for support, though Ukraine is not a NATO member.[32] These small numbers of troops used to be called "tripwires" because they would set off an alarm when overrun by superior numbers. We don't want to give the Russians any reason to increase their forces or attack our forces in these nations. Our senior congressional leaders talk like the "situation" in the Ukraine has been resolved. In Crimea the Russians have won. Putin has stopped his overt aggression in eastern Ukraine, so the United States and its allies are victorious. But all Putin has been forced to do is postpone his takeover of the eastern Ukraine. It is still in play.

Sanctions against Russia have highlighted limited cooperation between Russian and U.S. businesses in space and military hardware. The International Space Station gets its human resources on Russian rockets. They are paid some $60 million for each person they send into space. Russia wants to suspend operations of its eleven Global Positioning System ground sites and open negotiations with the United States for location of some of its own Glonass ground sites in the United States to do similar geo-location.[33] Glonass has a satellite constellation as large as GPS, but is positioned for better service in the Northern Hemisphere.[34] We might think about this for 15 seconds, but it is not going to happen.

Rosoboronexport, Russia's largest arms export company, is sharing military arms with Syria, suspected in helping Iran build its military capabilities, and selling helicopter parts to the allies in Afghanistan. Syria and Sudan sell arms to other countries that are exported or licensed by China. Rosoboronexport exports 80 percent of Russia's military hardware.[35] The U.S. Army also bought Russian MI-17 helicopters from a company in the UAE, using a U.S. defense contractor, ARINC. The helicopters were commercial versions, bought from Rosoboronexport, converted to military use before being shipped to Afghanistan.[36]

The business arrangements of arms dealers, their customers, and governments are constantly at odds, and occasionally in conflict; arms dealers don't always operate in the best interests of the governments that host them. We could ask if it is a good idea for the U.S. Army to buy Russian helicopters for the Afghanis, but we probably would not get an answer we want to hear. Apparently the arrangements have nothing to do with whether the country producing the arms is a friend, the company is operating under sanctions, or the United States and several other allied countries make helicopters that would work just as well. In the business of war, there are few enemies.

The Ukrainians and Russians have gone out of their way to not give any reason for the other side to start a larger war. Putin says there is no war; he is not involved. There are Russian nationalists in that part of the world, and non–Russian-speaking residents are threatening them. This is thin on logic, and by August 2014 it was starting to look incredible. After all, there are English-speaking residents in all parts of the world, but nobody claims their allegiance because of it.

Putin offers to meet with our Secretary of State, then our President, to start a dialogue. He calls the U.S. President to talk. He wants to avoid war, but he hears these stories about Russian-speaking residents being at risk. Whose intelligence is being insulted? Are both sides insulting intelligence, or just one? We have to think about that.

The second effect of the transition to cyberwar is that leaders in warfare are shifting emphasis away from kinetic war to a boa constrictor squeeze that requires more preparation, less bravado, and a more technical approach to winning in battle. It is total war. That means the orientation of politics, business, and armed forces to win, not necessarily territory, but the will of the enemy population. It was not a coincidence that oil and natural gas facilities were the first ones seized. Crimea depended on the Ukraine for both. There could be no concern for either by citizens of Crimea if the Russians intended to keep the people of Crimea on their side. The Russians, who cut off oil supplies regularly to influence the will of those affected, understand the concept.

This is not the kind of war that will be won with tanks and long-range bombers, and it is not a war that Western democracies seem able to fight. At a simple level, cyberwar means both controlling and using information to manage the will of the people. Cyberwar is how that gets done. The people are domestic audiences and foreign ones. Some think we are above this kind of manipulation, believing that ideas are sent around for people to digest and accept, or reject, as a matter of free choice. That is naïve.

In some countries, governments control the distribution mechanisms, modify content to suit their own needs, remove content contrary to their own ways of thinking, or allow access to content and then monitor those who have it. They manage thought leaders, sometimes with rewards for publishing what they want, other times through threats of assassination. The free exchange of ideas is not equal in the world—not even in the *free* world.

It gets harder to control as communications bypass governments and go directly from one person to another. Governments find it more difficult to track down single individuals in millions of e-mails, tweets, and postings on Facebook, but it is something they are capable of, given improvements in monitoring technology. Cyberwar is partly the use of this monitoring technology.

Vladimir Putin learned a valuable lesson in his run for election in 2012: social media can bite. It was a factor his allies did a poor job of controlling, allowing his main opposition candidate to get 17 percent of the vote, and reducing his total to 60 percent, considerably less than he anticipated. A BBC reporter followed voters around who appeared to be voting in more than one place and filing "absentee ballots." It was called "carousel voting" by those who observed it. The reporter was forbidden to photograph the voters, but she did so anyway.[37] This is politics, as practiced almost anywhere in the world, and we can bet it was practiced in Crimea. In the larger picture the extension of politics-as-usual outside the borders of the country is more than meets the eye.

These days a government has to be a man-in-the-middle to discover what information is moving through the media, molding the information to meet its needs, and influencing those sharing the information by either persuasion, reward, or disincentive. This is winning the will of the people by controlling the message, not through historical methods like brainwashing or government-run media. This is monitoring a population. This is managing information.

Western governments tend to believe in the power of ideas. If we offer a range of options and allow people to choose for themselves, they will find their way. We believe that truth will set you free, an idea that is shared by many. The concept comes from the Bible and is quoted on the wall at the entrance to the CIA headquarters. The new kind of war is different in that it is fought in the dark, in places where truth is only as deep as a plausible denial.

The absence of war favors covert special forces missions.[38] The military unit that killed Osama bin Laden was covert, and would have been unacknowledged—but not forever. When the neighbors were standing around with cell phones taking pictures of helicopters hovering around the house next door, any possibility of an undetected, covert operation vanished. The special forces shot family members and bodyguards, entering the house to find the man they were looking for. They wore uniforms; they operated under military and White House control; and they were later identified, by the White House, as Seal Team Six.[39] We should not know that information, because it destroys one of the reasons for covert action—plausible deniability.

If an operation is unacknowledged, the operatives will look more like local people, have greater risk if discovered, and possess deniability. If the mission doesn't go well and the operatives are captured, there will be a long delay before they see sunlight again. "Russian nationalists" were killed on the road to Slaviansk; Russian leaders cannot claim them as their own, even as they make efforts to recover the bodies.[40] The *Financial Times* reports a memorial service in Moscow for special forces soldiers, carried across the Ukraine border in a "Code 200" truck, a symbol for repatriated dead soldiers.[41]

The most important concept to grasp is that of total war. Whatever the motivation (make the world Sunni Muslim, facilitate China's leadership role in the world, protect Israel from its avowed enemies, allow citizens of a country to elect the officials they choose, or whatever else we can make up as a justification for battering other countries into a belief that one form of governance is better than another), countries should see war as all-encompassing. It combines every element of information flow, using whatever resources are available and whatever methods are effective to achieve the result: to manipulate, manage, influence, and discredit other views, both internally and in other countries. This is the way of the new war—cyberwar.

The methods include more than just what we can see or read about. In fact, they are rarely done in the open. They can be covert; they can be unacknowledged military operations; they can be lethal operations outside of a combat zone, like drone strikes; they can be counterterrorism operations.[42]

Deniability is essential to covert operations, and without it, credibility will suffer. A denial doesn't have to be rational, logical, or believable by the whole world. It just has to be repeated often, said with conviction, and directed toward people who will believe it. The Ukraine government posted pictures purported to be of the Russian military, showing individuals who were said to be in operations in multiple locations over several years. The U.S. State Department supported the photos' authenticity. The BBC, however, said this proved very little, and the Russians denied everything.[43] Reasonable doubt is important to the perception; deniability is essential, but truth is not.

The third reason for the use of cyberwar is that war is being fought for the same reasons it has always been—religion, politics, or egos, large or bruised—only the scope is bigger. Rudolph Giuliani, who was mayor of New York City on 9-11, said we weren't looking for war, but it found us that day.[44] We haven't been paying attention to cyberwar. It is finding us more than we would like.

War today is global. When we said we fought two world wars, we were exaggerating. Only a few countries fought those wars, and most (including the United States, all of Latin and South America, Australia, most of Africa, and Canada) had only scratches on their territories. The new wars are different; they touch every part of the world. In cyberwar, our enemies think big.

The fourth reason cyberwar is used is that kinetic war has become outdated. In the kinetic war, the enemy was clear. It is almost the same as tanks rolling across the desert to Baghdad. The enemy was clearly identifiable, dug in and waiting. The Iraqi Army attacked, every now and again, to show their will to win, but didn't like the results. Casualties made these wars unpopular, just as they did in postwar Iraq and Afghanistan. But if casualties had an effect on will, neither side would want to continue fighting. It is how they are perceived that counts.

In war, truth is not important. It is all about perception. Press people arrested in the Ukraine become a cause célèbre with the Russian media while arrests of Ukrainian reporters by nationalists are ignored.[45] The part of the truth that we tell influences behavior. We can't win this kind of war by running around the countryside with pictures that show it clearly wasn't a wedding party that we killed with that streaking missile, given the AK-47s and bomb-making materials found at the site. We are far too reactive to assertions made by people who are not our friends.

The real reason we can't win a war just by telling the truth is more deceptive—the recipients of the message don't care. The people of Crimea didn't give a hoot that the green-uniformed troops spoke Russian, drove Russian trucks, and took over key military bases in their first actions. Putin took over, a popular election legitimized it, and the pictures flowing out of the Russian press were all of happy faces. These are the happy faces of a new kind of war.

2

A Cloudy Day in Kiev

On a clear Ukrainian day in December 1991, Leonid M. Kravchuk was enjoying the sunshine. He had just won 60 percent of the popular vote and was in power after a long-running and dangerous separation from the old Soviet Union. Kravchuk had grown up a Communist, rising to the position of Chairman of the Supreme Soviet before things started to come apart in Moscow. He was a smart man who saw what was coming, and he switched his allegiance to a separated Ukraine.[1] In the mercurial world of Ukrainian politics, it seemed a natural thing to do.

He still had the election results from 60 percent of the country that favored him; more important, he had the confidence of the 90 percent who voted for independence from a collapsing Soviet Union. That majority, and his own popular vote, was making him a happy President-elect. The United States had already recognized the results, but in August President Bush made a statement that rattled the legislators, saying the President Mikhail S. Gorbachev knew best how they should manage the transition. That did not sit well with the leadership. The press called it the *Chicken Kiev* speech.[2] The Secretary of State came to clarify the position of the United States, and by November, the policy was flipped around to favoring Ukraine's independence. It was not a coincidence that it took awhile to work out.

There were the usual problems with becoming a new country, and one unusual one that influenced Bush's reaction to independence—the nuclear weapons still stored on Ukrainian soil. The Ukrainians wanted those, the Russians wanted them back, and the world wanted them to be protected from terrorists and other countries that wanted a ready-made path to nuclear status. There were plenty of bad things that could come from having loose nuclear weapons out of control, so every country in the world wanted to help Ukraine with its problem resolution.

Boris Yeltsin was throwing his weight around in Moscow and trying to make things difficult for the Ukrainians to go their own way. The Ukraine needed Russia for gas, oil, and fuel rods for their nuclear reactors, so they had to play nice with the Bear to their east.³ They concluded an agreement that allowed the Russians to keep their fleet in the Black Sea, as well as some troops and planes. In 1992, Kravchuk foolishly said he might take over the Soviet Fleet in Crimea and put its forces under his command. Yeltsin threatened to intervene if any country (carefully, without naming one) should attempt to change the status of the military agreement for having the navy in Crimea.

The nukes were still under the control of Ukraine, Belarus, and Kazakhstan, when the United States thought they should have been returned to Russia. The Ukrainians promised to give them back, but attached conditions that complicated matters. Confusing the matter even more, as a precondition, the Ukraine called for reductions in Russian troops in Crimea.⁴ The waters were calmed by visits from the U.S. Secretary of State, and hosting a Ukrainian visit to Camp David with the President. Everybody was friendly after that. We can't tell what happened at those meetings, but we saw the results.

Over the next few years, Ukraine made friends with NATO without seeking membership. With U.S. help, the Ukrainians brokered a deal that moved 2,000 nuclear weapons out of the country.⁵ The world breathed a sigh of relief, and then wondered how they came to have so many. Now, the world is wondering how we could have allowed all those Russians to stay in Crimea.

Things were moving along nicely until two things happened. First, three years after the conclusion of the nuclear agreement, Kravchuk lost the election, even though he was the favorite; a succession of people then entered the leadership, some pro–Russian and some not. Politics in the Ukraine washed back and forth between them. Second, in 2000, a former KGB light colonel took over from Boris Yeltsin as the leader of Russia, and he was not a live-and-let-live kind of guy.

In 2004, Viktor Yushchenko and Yulia Tymoshenko become the popular faces of the leadership change, soon to be called the Orange Revolution. But while he ran for office, Viktor Yushchenko's face was literally changed forever.

During the election campaign of that year, Yushchenko ran against the sometimes pro–Russian Viktor Yanukovych,⁶ who won. There was quite a bit of evidence of voter fraud, intimidation, and corruption. The Supreme Court threw out the results, so there was another election scheduled for December. We can count on one hand the number of times any national election gets thrown out by a court of its own country. This was a rare event. Rarer still was Yanukovych's reaction—he threw out the Supreme Court and promised new elections.

In September, Yushchenko fell ill. He was no longer the handsome leader that he once was. His face was pockmarked with lesions, and he was not able to campaign as much as he would like. He went to a hospital in Austria to see what the cause might be. His doctors said he had been poisoned with a strange and familiar substance, dioxin, and it turned out to be the second highest level ever recorded in a human being.[7] The Russians were the usual suspects, but nobody knew for sure. Naturally, they denied it.

Canadian MP Irwin Cutler also claimed to have been poisoned in the same way, around the same time, in a Moscow hotel. Cutler claimed his Russian contacts said, "It was a mistake," which was not much consolation. To Cutler, it was an explanation—nothing more.

At the time, Ray Boisvert, formerly of the Canadian Security Intelligence Service, said poisoning was unusual, since the Federal Security Service of the Russian Federation FSB usually used cyberattacks to harass their enemies, not something like this.[8] In the 1990s and early 2000s, politicians in the Ukraine were finding out there was a new level to politics, but they didn't know this was only the beginning. These events were nowhere near the Russian takeover of the Crimea; that did not happen for another 10 years. The Russians were then only just starting to use some of the new techniques the Internet had given them.

In an ideal scenario, participants are not at war and do not want to appear to be at war so as to avoid arousing the defenses of the enemy. This has to be more than denial—that is, there has to be more to it than just saying they are not at war, when they obviously are. This is why credibility counts. When bullets start flying and military equipment engages in battles, it is too late to pretend there is no war. The Russians were pushing the limit in the Ukraine, without going over the line, until August 2014. Just a few tanks, with a few anti-aircraft guns to cover them, were all they wanted to engage. When the Ukrainian government pressed nationalists in July 2014, the Russians watched without directly intervening. After July 2014, they increased the number of heavy weapons winding down those narrow roads to the west of the Russian border.

War should look a little like nations defending themselves against potential adversaries, mostly by keeping up their ability to fight, without ever fighting. The Russians don't seem to take this lesson of the Chinese very seriously, though they say the right words. Tanks, missile launchers, and motorized personnel carriers make it harder for one's neighbors to believe. If you live in a neighborhood where the military equipment is flowing, you feel like you should do something, or else get out of the way.

Some military forces prefer a course of preventative action called deterrence, but that is not an idea that fits very well here. Deterrence implies an enemy that must be persuaded to not engage, usually out of fear that they will suffer negative consequences. The perceived consequences must be great enough to convince them not to take action in the first place. That was the Cold War (and partly the reason for all those nuclear weapons). Both sides built up arsenals of weapons, few of which were ever used. It is good for arms merchants and militaries, but it doesn't do much to convince human beings that it is not in their best interest to start a war.

If Russia is to win, it doesn't want the enemy to see any kind of threat that might cause them to buy arms and get prepared. It doesn't hurt to have a person like Viktor Yanukovych in the President's seat, either. He can cut military spending, decimate the armed forces, and make deterrence impossible.

Russia doesn't want to have enemies; it wants to be friends, or friends with disagreements, but never enemies. It wants to trade with their friends, share open communications, have military exchanges, allow travel across borders, have diplomats in frequent contact, do cultural exchanges, and allow other countries the flexibility to deviate from paths they think are best, while making it clear that they don't necessarily approve of those ways. Friends can always disagree. Russia is the State Department vision of a friendly government—most of the time. We can always do another "reset" with them if things don't go well.

The Russians, who know about information war from practicing it for decades, target people in political leadership and thought leadership positions. If they take a long-term view, they want to target a few people who will be dominant political or thought leaders. They want to watch them, and influence them.

In a forgotten dispute with Estonia in 2007, over the removal of a statue, *Bronze Soldier*, the Russians were accused of launching a cyberattack against the websites of the Estonian president, its parliament, almost all of their government ministries, their political parties, three of the country's six big news organizations, two of their biggest banks, and firms specializing in communications.[9] This was just a little tiff, but it gave NATO cause for concern. Estonia was a NATO member and this might be war, but they had no policy to fight in this kind of case. In addition, the cyberattack was organized and broader than they were used to. NATO treated it like it was a first use of cyberwar to influence a position in another government. However, this was familiar territory for Russia, and hardly their first try at it.

The Russians have to spy on their friends, just to be sure they aren't telling them one thing, and then acting another way. Spying is not a direct way of influence. Every spy, or mechanism of spying, brings back information that has to be analyzed. The analysis is given to military and political groups that take action on it. Occasionally, raw spying data is presented to the public as evidence of something or another, but not often. The Russians would say, "Everybody spies, and the U.S. is doing more of it than us." The truth of who is spying the most is not material to this argument, but we don't know the truth of either side of it. After all, we don't know how much spying Russia is doing.

To make spying efficient, there must be focus. In spying parlance, this is called work factor analysis. What can I get for × amount of money, over Y amount of time? Are we going to get something that is worth what it costs; is there a return on the investment that is more valuable than the investment? So, they focus their spying activity on a small group of people worth knowing about—thought leaders.

The Russians had a balance to maintain in the Ukraine. Their nationalists broke into a "number of network nodes" in the Ukrtelecom, a Ukrainian telecommunications service provider, before Russian forces took over the Crimea and disrupted service by damaging the fiber cables.[10] This mostly affected telephone and other subscriber services, and it can be done for a couple of reasons. One reason is to disrupt communications. The other is to force communications onto circuits the disrupters control. The BBC speculates the Russians may have shown restraint in how far they went to disrupt communications because it is a double-edged sword: they find it difficult to monitor systems that are not working properly.[11] Their restraint may show as much about their capability to collect intelligence about Ukraine internal operations as it does about their cyberwar capabilities.

Their targets might be leaders in the world, somebody like the Dalai Lama, who has thought leadership over people in many different countries, or they can be a leader of a big automotive union in Germany, Italy or Spain. The Chinese target the Dalai Lama, stealing his personal letters and plans for the future.[12] They use them to steer and manipulate others who might cooperate or support him. It isn't a particularly new idea, but one the Russians are also good at.

Over the past 20 years, a relatively small number of books have been published about life in the intelligence services of many countries, but most of them are about the CIA and KGB. These two protagonists are the best known, largely because they choose to be. We also get the occasional book on the Mossad, the Israeli intelligence service, and China's Ministry of State Security, which is relatively new by comparison.[13] It isn't that the Chinese have not been spying for a long time; they just hide their institutions that do so. All in all, that is the way intelligence services should operate.

By all accounts, the Americans and Russians have been trying to influence each other's politics for a long time. Given their history, there is no reason to believe they stopped when Putin came to power. Given his background, he might even have been involved in some of it, before he got into his leadership position.

In 1968, for example, the Soviets offered help to Hubert Humphrey's campaign to try to keep Richard Nixon out of office. Nixon, who was anti-Communist, eventually became one of the best leaders for smoothing out relations between the two countries, but they could not have foretold that.

2. A Cloudy Day in Kiev

The KGB, the predecessor of the FSB, focused on the Jewish lobby, which they felt was too great an influence on the United States. They targeted "Scoop" Jackson, Leonid Brzezinski, and Richard Perle.[14] Some of the attacks against these men were total fabrications, but some came from legitimate sources and political muckrakers. The Russians have been practicing this kind of thing for a long time, but their techniques have changed as they went along. Now they focus on those who influence, with the Internet, particularly social media, as the medium.

A thought leader is a person who has expertise sought by others, and who leads by example. Some examples of thought leaders are Richard Branson of Virgin Group and Sheryl Sandburg, Facebook's Operations Chief. These are people who present novel business solutions to problems others share.[15] There are many areas where a person might be a thought leader, and it is not possible to find out who is most important without doing some analysis. It takes awhile to determine the thought leaders in particular areas of interest to people in a target country.

Political leaders are easier to identify, but occasionally more transient. They are elected or come to power in a country and have the power of their office, plus influence over others. Political leaders are actually easy to target, so the Russians would start with them, as they did with those around Nixon.

When the Russians targeted the Ukraine, they started by identifying key people and determining what they were likely to do in a given circumstance, within certain ranges. They want to find out what a few politicians have written and made public, maybe in campaign speeches, work environment, or personal life. Perhaps they can find blogs, a website, and so forth. They put a young officer to work on those aspects of life, and move on. That part is easy. Influencing groups through mass media is harder. Just ask any major advertising firm.

They also want to know whom these key people communicate with because they are the ones who have influence. The principle is the same in advertising, as when a company selects athletes to represent its products. People see the athletes wearing their tennis shoes or drinking a soda. Advertisers believe those viewers will buy their products as a result. That kind of persuasion is not always a sure thing, however. Athletes get injured; they hurt themselves with social media outbursts; or they might engage in some behavior like wife beating, stealing, or child molestation that makes them unattractive to potential buyers. Politics is much the same and just as unpredictable.

The Russians can see how to monitor influential politicians by watching the Chinese in major U.S. elections. The Chinese hacked accounts of the McCain and Obama elections teams, apparently looking for position papers and the directions that the candidates would take. The Russians likely assumed they were doing the same thing to the Romney team. Maybe they can share with the Chinese, or duplicate their efforts so the Chinese don't know they are hacking campaigns too. The Russians and Chinese want to know who influences the President so they can influence those people. If one of the influential people uses the Internet, they will know what they talk about and to whom. The Ukraine is small, so it is easier there.

All spies are fortunate that politicians in every country are arrogant (or careless) enough to ignore their security services and use personal phones, tablets, and mail. Hilary Clinton has been criticized for using her own personal e-mail system while Secretary of State. As any good intelligence person will say, that is tantamount to allowing the Russians, Chinese, the French and Israel's the opportunity to read your e-mail. They may not be the only ones, but they are good at it already and we don't have to guess about their capabilities. Angela Merkel complained about others monitoring her cell phone, which was her own personal phone and not an encrypted government appliance. She can complain all she wants but she should expect that countries don't allow an opportunity like that created by leaders who should know better, from slipping through their fingers. They are also smart enough not to talk about what they get from that kind of monitoring and would never admit doing it. These are rarely secure enough to stop a state-sponsored hacker, thus inviting other countries to look at or listen to what they talk about in their communications. The Russians especially don't believe that the NSA was the only one monitoring cell phones of leaders in other countries. There are probably no unprotected cell phones that aren't monitored by more than one spy agency.

So, the Russians collect a little information from sources who know their targets in the Ukraine well and from the Internet. They get access to their Facebook and Twitter accounts, as well as the target's Internet Service Provider (ISP) to determine where their e-mail is going. Getting cell-phone records in the hands of another country can be difficult unless they are using Russian services. In the Ukraine and Crimea, as luck would have it, some influential people are using Russian cell phones and networks, because Russian phone companies sell there, and they have network nodes where Russia can get access. They should; they built most of them.

Some targets are not using Russian services, so the spies have to work a little harder to get those. The security services might have to use some special tools to gain access to non–Russian phone services and ISPs. Telephone records are used to create maps of social contacts, the frequency of the contact, and whom the contact is with. U.S. law enforcement uses the same techniques to identify gang members, business fronts for money laundering, and cells of potential terrorists. Phone records are easy to get, if one can hire a casual hacker; in Russia, they are a dime a dozen.

The Russians want to determine who the Ukrainian gatekeepers are.

2. A Cloudy Day in Kiev

They want to be able to identify the person who answers the phone and mail, sorts, prioritizes, and forwards what needs to be read (because politicians rarely manage their own communications). They monitor the gatekeeper's work and home e-mail, messaging, and chat sessions. There are a few other hacker tools that will help with that task, sorting and prioritizing the sources. Other commercial tools map communications with various persons for frequency, type, and volume. They plug all their stolen communications into one of these tools and, where clusters come together, there may be thought leaders for the original target leader. The newly identified thought leaders are targets that will be monitored. Before long, monitoring the network of influential people becomes a significant amount of work, requiring a lot of resources, and this drives intelligence budgets. It is important to know what thought leaders influence a government either directly or indirectly; it is worth the money.

However, knowing what a person says will not be enough. Intelligence analysts need to be able to predict what those with influence will do, and then influence their decisions and message. Predicting actions or messages and influencing those actions and messages is much harder than collecting information. Our friend in the Presidency might tell all of his co-workers that he goes to church every Sunday, but only if someone checks will they know that he never goes during football season, or in years when the World Cup is held. At those times, he goes to the games on Sunday. Maybe he has a girlfriend his wife does not know about. These are not crimes, but can be used, if needed, to provide negative reinforcement, or leverage, in voting. This has a name, political blackmail, descriptive of its intent.

Intelligence analysts have to read e-mail and listen to phone calls to find out what the targets talk about. This is content—the what—versus with whom (their thought leaders). They probably can't pay attention to *everything* the target does or exchanges, so they narrow it down to topics related to, say, eastern Ukraine. They record everything they get from the Ukraine and sort on that term. Who does the target talk to about a topic? What options do they discuss? Who favors the position of not starting a fight over a few nationalist special forces in the neighborhood?

To be efficient in controlling the message, communications must be controlled, including the press. To make their job easier in the Ukraine, the Russians started with improving telecommunications in the country. This can be thought of as a "humanitarian gesture" with political benefits. It raised Ukrainians' sense of belonging to the country and market forces did the rest. They could then sell things to their targets—at least some of their phones.

Providing telecommunications infrastructure had the added side effect of connecting more people to Russian-controlled radio and television stations and making it easier to monitor target e-mail and phone. The Russians had to be careful to use their own services for secure communications, and not a provider from the European Union or United States.[16] In 2014, Dmitry Medvedev made a speech in Moscow to justify the upswing in Russian telecommunications infrastructure:

> We must have Rostelecom and its subsidiaries come to Crimea as soon as possible. We cannot tolerate a situation in which sensitive information and documents related to the administration of the two constituent entities of the Russian Federation are relayed by foreign telecommunications companies. This must be terminated.[17]

Parts of the world thought Medvedev was referring to NSA's monitoring of telephones exposed by Edward Snowden, but the ongoing business of Russian-controlled communications was proceeding long before anyone ever heard Snowden's name. The Russians have been doing the same things as long as any other country.

In 2011, Russia started the Europe-Persian Gateway, a high-speed fiber-optic network linking Iran, Russia, Germany, Austria, the Czech Republic, and the Ukraine. It runs across Eastern Europe from Frankfurt to Russia and Iran, all the way to the Sultantate of Oman, Muscat, and it has a designed capacity of 3.2 Tbps (terabits per second). It has both submarine cables and terrestrial transmission. The partners on this venture are Vodaphone, Cable and Wireless, Omantel, Rostelecom, and Iran's Telecommunications Infrastructure Company.[18] If the Ukraine collapses to European influence, some of the major nodes in their communications infrastructure will be controlled by Russia and Iran.

The new DREAM (Diverse Route for European and Asian Markets) fiber-optic route being put in by the conglomerate's carrier, MegaFon, puts Russia and Kazakhstan as a gateway for traffic going to Europe (Germany and Austria), the Ukraine, Slovakia, and China.[19] Tele2 Russia began to install LTE cell-phone service within days of the invasion.[20] No company works that fast, unless they know what is coming. Russia was trying to get a firm hold on telecommunications in the former Soviet bloc country. From the outside looking in, it was hard to miss the Ukrainian connections in their build-out.

People without an Internet or cell-phone connection have to be reached in other ways. The billboards in Crimea were made for an audience of people, two-thirds of whom do not have Internet access. Over half of Russians have it, more by percentage than China.[21] Absent the Internet, television is the weapon of choice; if there is little of that, then billboards and radio will do.

Next, the Russians want to be friends with the target's thought leaders who favor not starting a fight. They give them money through business friends; they provide assistance in writing articles about their positions. Magazines, radio stations, and television bring the thought leaders in to talk to audiences. (Almost all households in the Ukraine have televisions.)

The Russians have a Minister of Trade call a few politicians in the targeted group to seek their opinions on how they might be better friends in the future. They increase trade in the part of the country where the politicians live. The standard of living improves modestly, but not a lot—just enough to recognize the value of their trade with Mother Russia.

The intelligence analysts have their favorite reporters do stories on the candidates' good works. They distribute these stories through different outlets so they don't bear the byline of the same reporters, nor the same press agency feeds.[22] We have another good example of how this is done in Pakistan.[23] The *New York Times* carried a story about the mess a news agency can get into when it clashes with an intelligence service like Pakistan's ISI, known for its lack of subtlety and friendship with the Taliban.

Geo News, a popular CNN-like news channel, blamed ISI for a shooting attack on one of its reporters. The military put some heat on cable television operators to drop the station, which many of them did. Advertising revenue dropped off, four vehicles were burned in different cities, and someone beat up a journalist, calling him a "traitor." Pakistan suspended Geo's license for 15 days and fined them $100,000. The current situation came from having a popular news anchor, Hamid Mir, beaten up. His relatives blamed ISI, and their claims were broadcast on network news. The government of Pakistan sided with Geo, but allowed the suspension of the license.

The Russians limit political demonstrations, or "problems," in their political friends' region. They strike contracts for oil and gas that are popular. They tell their friends what wonderful folks they are and seek their help to identify additional ways in which assistance can be provided. They start the equivalent of a political action committee, supporting the political friend for a higher office.

There is to be an election and one of their friends wants to be President, while one of their enemies has managed to get support from quite a few people. The opponent is attractive and a good public speaker. The movement has a catchy phrase, such as "Orange Revolution," which is spreading faster than the Russians would like. Russia has spent a lot of time and money and doesn't want it all to go down the toilet because bunches of young people, over whom they have no influence, are trying to get elected. To the Russians, this doesn't look good.

The Russians covertly put more money into negative ads, and start to work on the election committees where they have some influence. They can bribe some people easily. They can work on the ballot process and try to manage it so their friend gets his votes recorded and some of the enemy's do not. They stuff ballots in parliamentary elections to strengthen their position.[24] They line up people to control certain areas where there is Russian sympathy. A certain number will support them, but they have options for ones who don't. They can bribe voters, or intimidate them if bribery doesn't work. They get people to vote multiple times. They try to get a contract to manage the voting machines in areas they control, and if that doesn't work, they hire a hacker or two to attack the voting system itself, changing the results. For example, CyberBerkut, a group of pro–Russia hackers, infiltrated the Ukraine's central election computers and deleted key files, rendering the vote-tallying system inoperable.[25]

In the case of the Ukraine, this novel approach almost worked. They managed to get favorable results for a couple of candidates, who were right-wing nationalists, but they faltered when discovered. The Russians tried to place people friendly to them as "voting observers" who reported irregularities in those areas where the supported candidates were not likely to do well. Their influence was winning out until the Supreme Court overturned the election results, proving bad things do happen to good people.

There will inevitably be people involved who not only don't favor the Russian position but also are violently opposed. As stated earlier, on the bell-shaped curve of opinions, a few people will favor a position, and a few will oppose the position. There may not be many who oppose the position, but they must be influenced as well. By reading the opposition's communications, the Russians are bound to discover things that will not look good to other political leaders and countrymen. The negative information can then be fed to a newspaper that normally favors the opponent. This is not a new tactic. The Russians pulled several dirty tricks like this prior to Reagan's second term in an attempt to keep him out of office. Yuri Andropov, the Soviet leader, was former KGB.[26] The negative information had credibility because these outlets were friendly to Reagan. They published the stories and Reagan was on the defensive for a month afterward.

The Russians know their opponent's travel plans—where he will be speaking and where he will be meeting in the coming weeks before the election. They make sure there are mothers of soldiers and concerned citizens present. In some cases, the same people are shuffled from one place to another. If they want to confront the opponent, or make sure his speaking engagements are not well attended, it is easier to take care of if they know where he will be.

The Russians can get a list of the opponent's major supporters and work on some of the bigger ones. They can write e-mails and letters in the opponent's name. They can write letters to political creditors and complain about payment of debt, or claim the use of political funds for personal expenses. Since a lot of people write things for political leaders, it is hard to trace back to Russia.

The Russians have a few guys they recruited in Donetsk move into an office building and hold it for a few days. They move on later to the Donetsk Airport, which is not as big as Heathrow or Charles de Gaulle, but still worth having. This is an annoying interruption to government business, but it is also a symbol of the inability of certain leaders to keep civil order.

Citizens picket the opponent's campaign headquarters. This is an incremental process, but, over time, it will discredit this person and lower his influence. (It isn't as easy as it sounds, and it doesn't always work.) Ultimately,

2. A Cloudy Day in Kiev   33

the objective is to have the opponent dropped from the President's list of thought leaders. The Russians continue to track that.

If this all sounds like politics as usual, there is something dangerously wrong with politics. This is war, and governments are behind it. They are manipulating the elections of other countries. If you think that is acceptable for the Russians in the Ukraine, then we have to ask whether it is acceptable for the Russians or Chinese to do it in the United States or Great Britain. They have done it before, so we might consider why they would stop now. Where does an open and free election process end and political manipulation begin? Is it appropriate for political parties to play these tricks on one another, but not allow other countries to do the same thing?

A country manipulating elections in another country has to work hard to keep its operations secret. While this is going on, other countries want to do the same kinds of things. In the Ukraine, Russia needed to disrupt the other country's covert operations where it could identify them. Spies usually know who the other spies are, to some extent or another, and cooperate with those who are on the same side. But they also don't want the sponsors of the other spies to find out what they are up to. They spend a third of their limited resources on keeping their operations secret, and another third trying to undo things that other governments are doing. It is expensive work.

During the Orange Revolution, the Russians used oil as a weapon. This is part of economic warfare. It started with a "dispute over pricing of gas" and escalated when the Russians cut off Ukraine's supply for a time. BBC, in its Ukraine Timeline, says this happened three times. It is a little crude, and transparent, so they may have overdone it. The cut-off of gas has hurt their friendship with the Ukrainian populace, especially those with no heat in the middle of January.

Still, the Russians managed to get their candidate reelected, but it didn't last long because the following year Yanukovych lost again. The imperfect electorate is fickle, to say the least. Reagan won in a landslide in 1984, after the Russians had done as much as they could to stop him. They must have learned that working hard does not always accomplish their objective. This is politics, after all. Voters are not as easily swayed by propaganda as governments might wish, and candidates for public office frequently find they are not as convincing as they might like.

With their candidate back in office, the Russians had to undo a Kiev legacy—a deal with the European Union for trade; they had to do something to discourage that. There was an existing "free trade" agreement with the Ukraine that was started long before any of the trouble there. The Russians warned the Ukraine government that the agreement would allow goods made in Europe to flow freely across the Russian border without any tariffs being applied, impacting jobs and revenue in Russia.[27] Putin was clear that he meant

European goods were high quality and likely to affect the manufacturing of Russian goods, particularly in aviation, automobiles, and agriculture. We would have to think a little to remember a time when a world leader showed fear of opening their borders to better goods than they could make themselves.

Yanukovych said the deal was not good enough, and he intended to examine the arrangement with the EU more closely. He hedged this with a statement that he wanted the deal to go through, but on better terms. This was a thin white lie, but politicians are prone to this kind of lie and treat it as a temporary deviation. The Russians had to check his mail to make sure he was saying the same thing in private.

A small number of demonstrators said they were going to set up shop in Kiev to protest the delay in making the deal with the EU. The Russians asked a few sympathetic people to join these demonstrators so they could identify the leaders. The number got bigger and their guy took action to clear the streets, and handled the "terrorists" who set up this operation. This did not go well, and increased public support for the people in the square. A couple of young officers in the FSB Ukraine Group were in trouble for their lack of success. Intelligence officers have to see the future, and these two were not clairvoyant.

Other cities had similar gatherings and it became more difficult to call them all terrorists. Their guy looked like a bully, using tanks and armored personnel carriers on these nice people. There were also press people there who couldn't be influenced or controlled. A few were abducted, and some were threatened; others were killed. A news team was held by Russian separatists, who found they had committed no crime except working for an independent news service.[28] They started "restraining" some of them and roughing them up. It helped a little, and some of them went home.

Viktor Yanukovych wrote a letter to Putin asking for Russian troops to help stabilize the country. Their UN ambassador held up this letter to let the world know that it was real and not something made up.[29] (We might wonder why anyone would think it was made up.) The UN members looked perplexed. The ambassador outlined all the reasons for this kind of action being authorized, but the UN had heard all this before. Yawning was noted in the video of the proceedings.

Before things got worse, the Russians moved regular army troops and enough airplanes to win an air war within striking distance of the eastern part of the Ukraine. NATO showed pictures of them in the news outlets, and the Russians denied those troops were there. They pointed out that "troop exercises" were announced and they weren't hiding anything.

At the same time, they moved Yanukovych out of the country and protected him against the possibility of ending up in jail on some trumped-up charges; they knew how that could happen. Things went downhill, and the government let Yulia Tymoshenko, of Orange Revolution fame, out of jail, where Yanukovych had put her to begin with. She gave the mob in Kiev a great speech that reminded everyone of why they were there, but was abandoned by the electorate. Politics are again fickle, and unpredictable.

There was a timetable in Russia for the operations that would lead to the takeover of the Crimea. The final stages of preparation were winding down. The actual takeover was about to begin. Now the military had to execute some real cyberwar activities. Russia started by jamming the cell phones of government leaders so they couldn't coordinate a response.[30] The Russian parliament announced its approval of the use of force in Crimea. Everybody knew what was coming after that. Before the Ukrainian leaders could get organized, the Russians activated a plan to move special operations forces into Crimea at key communications areas, oil and gas distribution points, and military bases occupied by the Kiev government.[31]

Russian forces were ready to establish their own power and telecommunications so they didn't have to depend on Ukraine. That started the following week.[32] They started putting up posters that brought the image of the Nazis back to families who still knew what that meant. In Donetsk, where paid thugs had taken the office buildings, they distributed flyers that said Jews must register, striking fear among many who remembered the Nazis at Babi Yar, in 1941.[33] The Soviet Union had saved them from the Nazis, and would save them from Europe this time.

The Russian forces took Crimea. A job worth doing, and one that took several years to prepare for. A war won, without a shot being fired in anger. It isn't over yet.

The Russians seem to take their lead from the Chinese, but they are using a club where a small poke with a finger would do. That club is energy. Gazprom, the monopoly oil company owned by the Russian government, is playing numbers games with the Ukraine. In March 2014, it said the Ukraine owed slightly less than $2 billion that had been carried as a debt. In December 2013, Putin made a deal with Yanukovych, the Ukraine leader, to lower the price and wipe away a substantial part of the debt, about $15 billion. Gazprom then paid Ukraine, in advance, its transit fees for moving the goods across their territory. When Yanukovych was removed for his own safety, Gazprom started recalculating the debt.[34] In two months' time, that had suddenly jumped to $22 billion. Apparently, it was all in how—and when—that amount was calculated. This huge debt made the currency value unstable and loans harder to get. The Ukrainians needed $1.6B just to buy what they needed for the coming winter, and they had trouble getting it together.

Europe, which gets some of that gas too, was not happy to have supplies threatened again. They knew this was economic warfare, and had some things they could do to leverage the Russian position. There is one thing about the Russians that is different from the Chinese when it comes to war: The Chinese learn from their mistakes and adapt quickly. They are patient to a fault. They don't like a fuss, and they have patience enough to not make one. The Russians have never been like that.

Shortly after the Crimea takeover, the United States and European Union announced the first sanctions against Russia. The second set of sanctions was even more confusing. The United States put sanctions on some individuals that Europe did not, and Europe sanctioned ones the United States did not.[35] In August, the situation changed, with Russian forces no longer pretending to be anything other than what they were. Mr. Medvedev, Putin's deputy, said the sanctions would be met with a ban on over-flights of Russia's airspace. That would prove interesting to enforce and lead to other countries wanting to do the same thing. World travel would be more complicated if the bans caught on. Finland was already wavering because it did so much trade with Russia, and would be hard pressed to fly away from Russia on every venture outside its borders.

Europe decided to give the Russians a week before the next set of sanctions kicked in, but promised to make them good ones. To most of us, they sounded like a tap on the collective wrist of the Kremlin leaders, but we should take a closer look, because economic warfare is not as simple as it seems.

The United States and European Union got practical experience with sanctions through dealing with Iran over several years. Iran is far from subtle. Iran's sanctions were popular because the European Union, and a few others, thought the Iranians were building a nuclear weapon, scaring the devil out of everyone in the Middle East. That was sufficient incentive to make sanctions work, though even that incentive was not enough to stop some violators.

The Iran sanctions had three components:

- First, prohibitions on oil, tankers, and the insurance industry made converting oil into revenue much more difficult, costing Iran billions of dollars.
- Second, banking sanctions cut off Iran from the international banking system, making currency conversion and trade more difficult.
- Third, the United States invoked the little-known U.S. International Emergency Economic Powers Act, which allowed it to block Iranian assets subject to U.S. jurisdiction. At the same time, the act made it illegal for U.S. citizens to do business with designated persons or companies.[36]

The State of New York brought charges against BNP Paribas, a French-owned bank, for violating the banking part of these sanctions. Under orders "from high levels of the Bank's group management," BNP Paribas engaged in a "systematic practice ... of removing or omitting Sudanese, Iranian or Cuban information" from U.S. dollar–denominated transactions, with the purpose of avoiding disclosure "to any potential investigatory authorities," according to a DFS (New York Department of Financial Services) document.[37] The $8.97 billion settlement is a lot of money, but the implications of this case went further than just the dollar amount. There was evidence that people in the U.S. bank were intentionally kept out of the loop on these transactions, and knew how widespread the practices of BNP Paribas were in other overseas banks. Banks and other financial institutions get indicted now and again, but they continue to violate sanctions in ways that are not traceable to them. Some international banks will, no doubt, follow the same kinds of practices in dealing with the oligarchs in Russia.

In the United States, the Department of the Treasury runs most of the sanctions against Iran, not the Defense Department. This is also true of the sanctions against Russia. Indictments are brought by the U.S. Justice Department, or, in the BNP Paribas case, a state government, and the settlements are generally to avoid criminal prosecution. Trial in open court is an event they want to avoid.

The first recognizable targets of sanctions in Russia were eleven friends of Putin and a bank, Bank Rossiya, which was said to be used by some of the very people named: Yury Kovalchuk, Vladimir Yakunin, and Andrey Fursenko. Kovalchuk owned 38 percent of the bank, among other things.[38]

This certainly was not going to set Russia back very much, since they have more than one bank, and 15 larger ones than Bank Rossiya. The bank told its main customers that, given the sanctions, it would do all of its business in rubles. With the value of the ruble falling, only partly due to Crimea, this created some risk for Russian banks. They were also exposed in Ukraine by having $28B in loans there.[39] Those loans are worth less, and the political climate may make them harder to collect.

A second round of sanctions began in July 2014, and there is a good chance more will follow. This round started with OAO Rosneft, the state-controlled oil company.[40] The sanctions also named OAO Novatek, Russia's second largest gas company; OAO Friends of Putin bank, the bank connected to the gas-export monopoly; and Vnesheconombank (a.k.a. VEB), a state-owned development lender that was backing most of the Sochi Olympics construction. The sanctions included the Russian Agricultural Bank, the Bank of Moscow, Kalashnikob Concern, and munitions maker Bazalt. A state-run shipbuilder, United Shipbuilding Corporation, was created by decree of Vladimir Putin in 2007. Its home base is in St. Petersburg, where Putin used to live, and it is the largest shipbuilder in Russia. It owns half of Finland's Arctech Helsinki Shipyard.[41] The entire list of companies is contained in a five-page document, which includes various subsidiary institutions.[42] The Russian news service, RIA Novosti, published a story three days before the announcement of the sanctions, saying they were not affecting Russian banks. That was before they had even taken effect, or the list had been released.

Eventually these companies will find it difficult to find dollar financing for their business operations, forcing them to get their money from the Russian Central Bank. With the foreign-exchange reserves dropping by $100 billion in July, to prop up the ruble, the Russians will have a hard time finding a sympathetic ear for their statement that these new sanctions "will have no effect." They said the same thing about the first round of sanctions, and that isn't working out as well as they hoped.

Lee Wolosky, a well-known attorney in international business, says we would be better served by going after the oligarchs who support Putin, while mounting an information campaign showing public corruption in Russia, particularly among its leaders.[43] These are things that can be done without harming the average citizens of Russia, who have it hard enough as it is. Michael Weiss, of *The Interpreter*, says Putin and his oligarchs have Swiss bank accounts and we know where they are. Give the Russians 24 hours to get the tanks and missiles out of Ukraine, he argues, or make them public.[44] It is hard to see the motivation for the oligarchs to react to news like that, but it is not much different from some of the other economic sanctions levied against the Russians. Their public statements say that nothing seems to bother them.

We forget that not every country has the same relationship between business and government, and the differences are important to economic warfare. They affect how the allies can respond. Russia has a relatively small circle of billionaires and former FSB people running its government. They are friends of Putin, or they wouldn't last long. This limits the targets.

Here is the curious thing about all of what is described above: The Russians deny doing anything to anyone in the Ukraine. Their problem now is their loss of credibility for their statements. It must be deniable to be effective, and their actions are way past deniability. We should expect them to deny doing anything to manipulate, spy on, or disrupt the normal operations of anyone in the Ukraine.

By early September 2014, the Russians had put troops and heavy weapons into the Ukraine while Europe and the U.S. allies stood by.[45] President Obama nixed the idea that Ukraine might join NATO and did not want to send arms. The Ukrainians were then pushed back to their original positions by overwhelming force. Putin said he could be in Kiev in three days and that the region could be declared for statehood, but the Russian press "clarified" his comments. He didn't really mean it, they claimed, as a Russian-speaking woman held up a map of *New Russia*, taking a third of Ukraine. Do we think he really meant it?

In October, Russia asked Lithuania to help recover some "deserters" who failed to report to the Russian Army in the 1990s when they were part of the empire. This further limited travel of people who were wanted in Russia's new empire, and it reflected the position of Ukraine's leaders who were being defined as criminals. Now it all depends on which countries are willing to honor an extradition order, which changes with the wind. Lithuania is a current target of Russia, which has monitored individuals on their computer networks, just as they did in the Ukraine.

The range of things the Russians have done in Ukraine is largely defined by circumstantial evidence that would not hold up in court. Certainly, the Russian denials of any involvement there seem unrealistic at best. There aren't very many others we can blame for what seems to be a coordinated attack against a country's information infrastructure, and armed troops occupying territory. They look Russian, wear Russian uniforms, and are memorialized by citizens of Moscow. Eventually the Russians, who are impatient, will learn from their experiences. In the meantime, the Europeans are starting to pay attention.

This is the future of war. No more lining up in uniforms of just the right color and blowing each other to bits. This is war by deception and denial. A war not fought by militaries.

3

THE LAST GREAT WAR

> *War therefore is an act of violence intended to compel our opponent to fulfill our will[1] ... the compulsory submission of the enemy to our will is the ultimate object ... it requires the destruction of [the enemy's] military power, the conquering of its country, and the submission of the will of the enemy.[2]*
> —Carl von Clausewitz, 1832

Since 1945, we have not been involved in very many wars that resulted in the enemy submitting to the will of the victor. The Second World War probably worked out better in that regard than most that followed. The Japanese, Germans, and Italians stopped most of their aggressive actions against their neighbors, though it took two wars to get them to that point. If we measure war on the success or failure of a strategy to compel our enemies to bend to our will, then we have lost most of the wars we have been in since then. No other country has done much better.

It is more difficult to define war now than it was a few generations ago. If we start with the Uppsala University conflict database, we can quickly see that most of the current wars are in Africa, the Middle East, and Western Asia. We are in a period with reduced casualties from war, which might be a cause worth celebrating *if* it lasts.[3]

Uppsala uses a definition of warring parties favored by the United Nations: A *Warring Party* is a government of a state or any opposition organization that uses armed force to promote its position in the incompatibility in an intrastate or an interstate armed conflict. The criterion for determining activity is at least 25 battle-related deaths during the specific year in the dyad of the warring party. We almost have to take a breath after reading that description.

There are many nuances to how this is formulated. Some of the events characterized as occurring between warring parties may not seem like war, unless it happens outside your hotel room. There can be state or non-state actors in this play, but sponsorship of a state may not be recognizable by researchers, since the state will go to great lengths to see that nobody can trace that support back to them. Like the Russians, the researchers will have no idea where those armed men came from.

Iraq, in 2003, may someday be seen as the end of the Last Great War, fought first by the Iranians, then by the allies, and now by the Iranians again. It was an example of a military action that overwhelmed a leader and bent the country's will to that of the occupying power, at least long enough to replace that ruler with another (and, since then, with yet another). The Last Great War didn't end the way von Clausewitz thought it should, nor the way the allies dreamed it would. However, it did end the way of recent wars.

Iraq didn't fight the allies at the beginning of the Last Great War—only at the end.

In a curious set of circumstances, the Shah of Iran threw out Ruhollah Khomeini, who took refuge in Iraq, of all places, for fourteen years. The Shah's secret police cracked down on his followers, slowly eroding the support the government had, and the Shah was ultimately driven out to Egypt. In February 1979, Khomeini returned to Iran. In October the Shah became ill enough to require treatment for cancer. There were a number of reasons the United States accepted the Shah when he left Egypt, but Iran did not like any of them. It became the basis for deterioration in U.S.-Iranian relations that would last until the present time.[4]

During the deliberations in President Carter's office, Carter went around the room and asked if the Shah should be allowed to come to the United States for treatment. Everyone agreed. Then, as described by Vice President Mondale, he asked another question. He said, as if he had reason to know the answer, "And if [the Iranians] take our employees in our embassy hostage, then what would be your advice? And the room just fell dead silent. No one had an answer to that. Turns out, we never did."[5] The first experience with the new regime was not a good one, and it set the tone for our relations ever since. On the U.S. side, the capture of hostages taken at the American embassy was the epitome of radicalism.

When Khomeini returned to Iran, he wanted to establish an Islamic State, and he did it with the blood of his enemies.[6] He sold the idea that the United States was the Great Satan—the ultimate adversary. The French and British have since joined us in that category, which is no small comfort. Khomeini galvanized his public with the idea that the world was against the Iranians and he needed their support to beat back their enemies. In the end, he got what he wanted and attempted to spread his revolution into Kuwait, Saudi Arabia, and Bahrain, all ruled by Sunnis but with a population comprising, like Iran, majority Shiite Islamists. The Sunnis and Shiites fight over much of the Middle East, but that is more about power than about who is right in a religious difference of opinion. Religious extremists have one thing in common: they all believe that their religion should govern their personal behavior, their law, and their government. Christians, Jews, and Islamists alike share that ideal.

Watching this revolution was one Saddam Hussein, who had come to power in Iraq in 1979. Khomeini saw Hussein as a Sunni who oppressed the Shiite majority in his own country, and made no secret of his dislike for him.[7] Hussein saw the Islamic Republic as a threat, the country of Iran in disarray, and thought a preemptive strike would be a winning strategy. The Shatt Al-Arab waterway, running northwest from the Persian Gulf between the two countries, would make a good excuse for a fight. In September 1980, Iraq invaded Iran. They fought a brutal war for eight years, with neither side winning much of anything. It was as ugly as the U.S. Civil War and killed more noncombatants. Neither side has forgotten about this.

Iran had made the mistake of killing most of its high-ranking officers during the revolution.[8] Tactics and military leadership suffered as a result. The Kurds had also started a revolution and were fighting both Iran and Iraq, while at the same time trying to stay away from both of them. The Kurds haven't stopped fighting, having met with some success. The disarray in today's Iraq gives them ample opportunity to pursue their goal of an independent state, but they rely on the goodwill and cooperation of the people around them, especially Turkey, where an oil pipeline serves as their lifeline.[9] Iraq's central government still tries to keep the Kurds from selling that oil without going through them, so their first filled tanker sits and waits.

In 1980 Iran had to adjust its strategy to its capabilities. It sent thousands of "human wave" attackers at the Iraqi forces that had modern weapons. Both sides bombed civilian populations in major cities. Iraq used nerve agents and gas on military and civilian targets, including Persians and Kurds. There were at least a half a million confirmed casualties (less than in the U.S. Civil War), with some estimates as high as a million and a half.[10] In Iran, the influx of weapons on both sides left a bad feeling about the United States.

Both Iraq and Iran shot at merchant vessels and tankers passing oil into the Gulf, which was not very helpful. Countries that used that oil were upset by the prospect of delays in refining it. The United States and Russia thus stepped in to stabilize their friends. The United States helped Iraq; the Soviet Union helped Iran. Russia and the United States were in a proxy war, part of the old Cold War, but Iraq and Iran could hardly appreciate that when their people were dying.

At one point the United States provided intelligence to Iraq, knowing Hussein was prepared to use both sarin and mustard gas against the Iranians, but sure neither that the Iranians would detect it nor that the Iraqis would use it.[11] In the end, they did use the gas, and it helped end the fighting, resulting in a treaty. Ultimately, neither the United States nor Russia got what it wanted from the war. Iran and Iraq liked the outcome even less. Khomeini said the agreement was like drinking a cup of poison.

If there was doubt about Iraq having weapons of mass destruction, we could be sure the United States and the allies knew it did. At the end of the war, in 1988, both sides settled back to normal, but were drained of resources and the stomach for more fighting. The war ended in a draw, but the reasons for fighting did not.

In an interesting footnote to the war's end, Iraq started work on Project Babylon in 1988, intending to produce several large-caliber guns capable of launching projectiles up to 620 miles. Space Research Corporation, run by Gerald Bull, managed the project. In 1980, Bull had pleaded guilty to illegal arms shipments to South Africa. In 1990, after completing a good bit of the work on a new gun, he was shot dead while entering his Belgium apartment. In his pockets was $20,000, ruling out robbery as a motive.[12] The following month, UK customs seized the final eight sections of a 1000-mm gun barrel. The UN supervised destruction of the remaining components.[13]

Hussein was upgrading his army of 1.3 million men, many of whom would soon be unemployed, which, in dictatorships, is not a good thing. Idle hands make revolutions. Hussein started laying the groundwork for using these individuals. He blamed his neighbors for his economy, which was falling fast. Kuwait and the UAE were first among many on this list, because they were said to have caused the sharp drop in oil prices and the devaluation of Iraq's currency. By April, Hussein had made up his mind to do something with Kuwait and, if that was successful, the oil wells in the eastern part of Saudi Arabia.[14] In August, 100,000 troops with 700 tanks invaded Kuwait. Within three days, the country was his.

General Samarrai, Iraq's head of Intelligence, wondered why Iraq wasn't warned to stop during the buildup of forces along the southern border with Kuwait. He knew the United States and others in the region couldn't have missed it. It is hard to hide large numbers of tanks even with camouflage. The Iraqi leaders took this as a sign that the United States might not engage. Hussein was convinced the U.S. experience in Viet Nam would keep them from doing much, and the lack of any attempt to intervene at the start gave him reason to believe he was right. He was going into Saudi Arabia next; his plan seemed to be working.[15]

The response to the invasion was not swift. It points to a military shortcoming that involves logistics and political activity: coalition building. General Norman Schwarzkopf, military leader in theater, briefed President Bush at Camp David that, if the intent was to drive the Iraq army out of Kuwait, it would take 8–10 months. All he could do in the near term was defend Saudi

Arabia.[16] That was basically logistics. It was 2003 before the invasion of Kuwait finally resulted in dislodging Saddam Hussein. The delay was coalition building. Wars like this one do not happen quickly. Impatient people do not fight them.

If we think about this in the context of the Russians in Crimea, the Russians had to have been planning their invasion for many months before they could pull it off. They must have relied on the Western countries' inability to organize quickly and achieve a coalition of defenders. There was nothing spontaneous about it, and it would not have been related solely to the ouster of Viktor Yanukovych. The Russians were planning this move for a long time, and are probably planning more operations like it. They are pouring tanks and weapons into the east of the Ukraine, even after international attention gave them pause. They obviously don't care about what other countries are saying; they just care about the outcome.

It may be years before anything can be done that will affect the Russians' ability to extend their domain. If Europe or the United States is in a hurry, they will be disappointed. There was very little anyone in the West could do once the takeover of Crimea was set in motion because the logistics and coalition building take longer than an offensive action. We should have been thinking about it for a long time beforehand. The only ones not worried about those two things were the Russians.

When the final assault against Hussein took place, the world saw the awesome power of the old wars. What they didn't see were the initial attacks, signaling the beginning of the battle. In the first hours of D-Day, special operations helicopters and air force planes hit thirty-one Iraqi observation posts along the Saudi and Jordanian border. Commandoes at Ar'ar, Saudi Arabia, and special forces in Jordan spread across the border into western Iraq. Special operations forces from Kuwait, augmented by the CIA, took up positions to monitor the Highway 1 bridges, dams, and the southern cities.[17] These were covert operations that were hidden from the Iraqis and from the Western press.

It wasn't until two days later that anyone saw a sign of war they could recognize. When they did, they heard a new term: "shock and awe." The live news coverage of bombs falling on Baghdad could be seen and heard, almost felt, as clouds of fire surrounded buildings in the background. We saw an endless stream of videos, with memorable moments like dropping a laser-guided bomb down a ventilation shaft of the Iraqi Intelligence Headquarters. It seemed like the devastation was more than any country could endure.

Deadly airstrikes, Tomahawk missiles, massed tanks, mobile artillery, and troop carriers took down the Iraqi forces facing them within days of the opening shots. Night-time operations were the mainstay in the first hours. Night-vision devices on almost everything, from helicopters to tanks, made

3. The Last Great War

that practical. Stealth aircraft were hard for radar to see, and carried deadly accurate bombs.

The allies had embedded reporters going along to report on everything from life in a tank to what roles women should be assigned in combat areas. Mostly, they just showed Iraqi prisoners and bombed-out relics of the army. Anyone could see who was winning. This, too, was a battle won by preparation, coordination, and firepower.

Everyone remembers the "Highway of Death," with its blackened carcasses of vehicles, drivers and passengers burned in place at their posts. It was a reminder of the consequences of war, like in the movie *Gladiator* when General Maximus turns to his subordinate and says, "At my signal, unleash hell." We all know what's coming. We don't have to wonder which population those images were made for, nor who the winner would be. If Julius Caesar would have had television crews around, he might have avoided his own end.

By the time the UN was able to inspect Iraq's weapons, they found a few things they had expected to find:

> The Iraqi chemical weapons stockpile consisted of chemical warfare agents filled into munitions and bulk containers. Iraq declared an overall production of some 3,850 tons of chemical agents during the past chemical weapons program.
>
> Approximately 3,300 tons of mustard gas and nerve agents, Tabun and sarin, or a sarin/cyclo-sarin mixture were weaponized into about 130,000 munitions, out of which over 101,000 munitions were used during the Iran-Iraq war. The Iraqi chemical arsenal, produced before 1991, included the following delivery systems: 155-mm artillery projectiles, 122-mm rockets, missile warheads, and a variety of aerial bombs. While most of the weaponized agents were filled into aerial bombs, the 122-mm rockets and 155-mm artillery projectiles were the most numerous munitions of the Iraqi chemical weapons arsenal. Iraq declared (and inspectors confirmed) that the 155-mm projectiles had been filled with mustard gas, while the 122-mm rockets were weaponized with sarin or a sarin/cyclo-sarin mixture. Iraq also declared that it had successfully developed and tested a limited number of binary artillery systems, including 155-mm and 152-mm shells for sarin, but did not enter into serial production of such systems.

Many of the "facts," like the statements of a man code-named "Curveball" and letters originally said to be from Niger claiming that Iraq was seeking a reliable source of uranium, turned out to be faked.[18] Who faked them is another matter. Nobody is saying. By that time, the war was over for Saddam Hussein, though not for many others.

It wasn't long after the initial victory that snipers and improvised explosive devices took over. If we look at the situation in Iraq today, we might think the war never ended. Increasingly, wars are fought by something other than conventional military forces. They can be covert special forces or Russian nationalists, but to noncombatants it is hard to see the difference. Equally difficult in today's world is figuring out who those noncombatants are.

We used to have an idea that civilians were not part of war. We have had something called the Geneva Convention since 1949. It was even updated in 2005. To some degree, most governments appear to have supported this idea, at least until they don't have to. Some of the rules are simple:

Customary International Humanitarian Law, March 2005, Distinction between Civilians and Combatants

Rule 1. The parties to the conflict must at all times distinguish between civilians and combatants. Attacks may only be directed against combatants. Attacks must not be directed against civilians. [IAC/NIAC]

Rule 2. Acts or threats of violence the primary purpose of which is to spread terror among the civilian population are prohibited. [IAC/NIAC]

Rule 3. All members of the armed forces of a party to the conflict are combatants, except medical and religious personnel. [IAC]

Rule 4. The armed forces of a party to the conflict consist of all organized armed forces, groups and units that are under a command responsible to that party for the conduct of its subordinates. [IAC]

Rule 5. Civilians are persons who are not members of the armed forces. The civilian population comprises all persons who are civilians. [IAC/NIAC]

Rule 6. Civilians are protected against attack, unless and for such time as they take a direct part in hostilities. [IAC/NIAC]

Rule 7. The parties to the conflict must at all times distinguish between civilian objects and military objectives. Attacks may only be directed against military objectives. Attacks must not be directed against civilian objects. [IAC/NIAC]

There are actually four conventions: (1) the first covers wounded and sick soldiers during time of war; (2) the second addresses wounded, sick, and shipwrecked military personnel during war; (3) the third, prisoners of war; and (4) the fourth, civilians, including those in occupied territory. Not all countries have signed all four, but 194 have signed the majority of them (which doesn't seem to matter very much to the people who ignore them).

Contrast this with a fatwa, sanctioned by nobody in particular, but issued in 1998 by Osama bin Laden, his deputy al–Zawahiri, and three others, aimed at having Muslims kill Americans: "We do not have to differentiate between military or civilian. As far as we are concerned, they are all targets."[19] We should have known bad things were going to happen with that kind of approach.

Terrorists do not follow any rules of war, but somebody's terrorists are fighting most of the wars today. They kidnap noncombatants; intentionally kill noncombatants; torture noncombatants; kill political supporters of people in power; kill teachers of girls who want education; behead people in public; fly commercial airplanes into public buildings; blow up trains, buses and cars of noncombatants on their way to work; kill girls who advocate for the teaching of girls; and run criminal enterprises to support their operations.

In three successive days in March 2014, Iraq had thirty-one people killed by bombs exploding in cars and in a café. In February there were fifty-plus killed the same way. The bombs are in Shiite and Sunni areas, and they never seem to stop. The will of these people is to fight. When there is no persuading the other side to adopt your beliefs, fighting will follow.

As part of the debate over whether to establish a no-fly zone in Syria, Turkey argued for the United States to do it, so as to stop the Syrians from using their home as a base from which attack. One of the arguments at the time, besides the expense, was that we couldn't clearly figure out which groups to support. Some were Al Qaeda; some were radicals not affiliated directly with Al Qaeda; some were just people who wanted the Assad dictatorship ended. Turkey, Saudi Arabia, and Qatar didn't wait for anyone else to act. They set up a base in Turkey to funnel supplies to the rebels, funded by "private sources."[20]

Early in 2012, the U.S. President started a covert program to support the rebels, mainly the Free Syrian Army.[21] It isn't really an army, but rather a mash-up of several rebel organizations that were fighting Assad's government. It started as a nonsectarian organization, but was disrupted by internal disputes between Saudi and Qatari competition over leadership.[22]

One of these splinters was a group called ISIS, or ISIL, the Islamic State of Iraq and Levant, led by Awwad Ibrahim Ali al–Badri al Sammarai (a.k.a., Abu Bakr al–Baghdadi and/or Abu Dua), who had a similar job opportunity before coming to Syria.[23] He spent 2005–2009 as a prisoner in Bocca Camp, an allied camp in southern Iraq. He was freed and, in 2010, led Al Qaeda in Iraq, after its former leaders were killed by U.S. and Iraqi troops. These are really bad guys. When someone hangs out with them, everyone knows he is guilty of something terrible, though there may be doubts about exactly what that is. It is guilt by association, but justified for this group.

We found out later that Abu Dua formed ISIS into a group that was known for killing Shia Muslims, Christians, and anyone who opposed them. He was motivated, well funded, and nasty. Al Qaeda claimed ISIL was not part of its alliance, and the popular belief in Iraq is that Al Qaeda is not capable of mounting the same kind of operation. Al Qaeda went through a phase when anything was acceptable as long as it produced a victory against the enemy. They learned that killing too many innocents was not good for fund raising. Given the nature of the two organizations, however, the difference hardly matters.

In June 2014, ISIL left its home and started attacking Mosul, Iraq. Mosul was probably best known as the place where Hussein used poison gas to attack the Kurds right after the end of the Iran-Iraq war, and where his sons died in 2003. The residents of Mosul have no great love for the central government of Iraq. The casualties in the ISIL-Iraq army were not high by Iraq standards, but ISIL claims to have lined up the soldiers they could find from Iraq's army and executed them. They claimed it, and showed the video. Iraq's soldiers, who saw it, knew they had to fight to the death, or run, when these guys showed up. Several stories say Iraq's army dropped its weapons, changed clothes, and fled. "We didn't see anyone fire a shot," said one man.[24]

ISIL fighters accomplished their objective by spreading fear, and they moved to within a few miles of Baghdad. They were able to do this because they had the support of Sunnis in the parts of Iraq where their spread began. In August 2014, ISIL was uncomfortably close to the Kurdish Regional Government capital of Irbil, and a major center of oil. A photo tour on Google shows an almost new, under-construction-everywhere city that looks clean and sparkly. It is hard to imagine the ISIL troops were just over that ridgeline. The United States finally decided to attack them using carrier-based fighters and armed drones. ISIL was not so quick to advance after that. We discovered, in the following days, that the Iraq government had withheld funds for oil over an internal dispute, and the fighters in that region were not well armed. The Iraq government was falling apart, but finally sent ammunition to the only ones fighting ISIL. The Kurds seem to not be loved by anyone, but they can sure fight.

As the advances stopped under increasing U.S. airstrikes, the French pondered the money. That was because France knew ISIL was building an empire from funds gained in several ways: those "private donations from Middle East benefactors," protection rackets in areas it controlled, extortion from refugees leaving those areas, ransom for hostages, sales of oil in areas it controlled, and a bank job in Mosul that was said to be $430 million.[25] (The bank job was denied by almost everyone, including the banks in Mosul, and considering the source, Ahmed Chalabi, the man who convinced the Bush Administration that Iraq had weapons of mass destruction, it probably didn't happen.[26]) The French reminded U.S. leaders that the operation in New York cost a million dollars to conduct, and ISIL had billions, so $500 million, here or there, was not going to matter very much.

All over the Middle East, and increasingly in Africa, wars are fought between Islamic factions and between Muslims and non–Muslims, but we might think of Lenin, rather than the current leaders, before offering an explanation. Religion may well be part of the reason these wars are fought, but it may not be the reason for the fighting. War is mostly about power.

Lenin thought the French Revolution took the right approach. The goal was to eliminate the landowners and ruling class and replace them with people "who can see that government and war are part of the same thing." In war, some government is always involved, and they are usually involved in ways they won't admit. Somebody is behind ISIL, but we have yet to look for who that might be.

In Iraq, the war goes on, absent the allies, and absent Saddam Hussein. Groups like ISIL make sure the leadership change, at least in areas they control, is permanent. Within the will of the people, the reasons for fighting are still there; just the government has changed. The Shia will not let that stand, so they are going to be fighting this war for a while.

Wars are becoming longer affairs. They take time, preparation, and planning. Anyone who thinks the likes of ISIL can just decide, in one day, to leave their captured territories and swing down into Iraq, is not thinking clearly. ISIL, like the Russians in the Ukraine, has been working its magic for years.

4

The War Peeks Out

If there ever was a defining day for the new kind of war, it was 17 July 2014. On that day, two seemingly unrelated events did more to show the nature of modern war than most others. The first was the shooting down of a commercial airliner in Eastern Ukraine, and the second was the Israeli invasion of Gaza.

The Russians had denied involvement in anything related to the fighting in the Ukraine until tanks and missile launchers started coming across the Russian border into the Ukrainian countryside. They are too big to hide, and NATO had been saying for months that they were being loaded onto trucks in Russia and moved over the border. Driving vehicles like these is a mechanical skill, but firing them is harder. Long-range missile launchers, like the SA-11s, have missiles that reach out to 70,000 feet or more. They are not easy to use properly and rely on defense networks, to help them. The Russian nationalists didn't have defense networks, so they didn't see that the targeted airplane was on a commercial flight path followed every day.

The first, and most notable, event in the two theaters did not require intelligence collection and analysis. A passenger plane was shot down in a location where the Russians couldn't control the press or the flow of information from locals. Nationalists controlled the immediate area, but it was part of the Ukraine. They met the accident investigation personnel and limited what they could see. One person interviewed later called the nationalists' efforts to control them "comical," but it didn't seem funny on the video. The "thug" description for these guys seemed apropos. The other event took place in an urban area where people siding with Hamas put themselves in harm's way to further the cause and embarrass the Israelis into stopping the fight.

In the Ukraine, Cable News Network (CNN) was better than the intelligence services for current, credible information that won the will of the people. CNN got the news out days before the U.S. government gave its assessment, almost as fast as information issued by the Ukrainians. There were videos of debris falling from the sky, taken by a freelance journalist, and of a surface-to-air missile launcher traveling up a road in the neighborhood the day before. CNN obtained intercepted voice conversations between the nationalists who shot down the plane, indicating the moment they recognized their mistake. It was not a cargo plane, but a Malaysian airliner with 295 people on board. The Russian language, peppered with swear words when the locals figured out what had happened, was blanked out in the translation, and bleeped for those who might speak the language. Once translated, the salty language was removed. Body parts were blurred, but recognizable as something once human.

CNN had a cell-phone video of the SA-11, minus some of its missiles, on the back of a truck being carted off from the scene. They had a video of a Russian officer, alleged to be the same one in a recorded conversation at a wedding where nationalists had entertained him. They had analysis by a former inspector general of the Transportation Department postulating that the Russians had the black box from the airplane, and reminding people that they did the same thing in 1978, when they shot down KAL 007. At that time, the Russians had denied finding the crash site of 007 and denied having the black box. It took ten years to find out the truth. In 2014, most of us had already forgotten about KAL 007.

By the second day after the disaster, CNN had enough to prove to anyone listening that the Russians had started a fight that got out of control.[1] It had reports from the crash site and the road that the SA-11 used to exit the area. CNN had collected enough evidence to place the blame. Europe would be listening, and any reluctance to apply more sanctions might go out the window.

Two days after CNN finished its local coverage of events, the U.S. intelligence community held a press conference to share the evidence of intercepted voice conversations, missile tracks, and debris fields. It was far too late to say more than "We saw that too."

Putin's first reaction was to blame the situation in the eastern part of Ukraine on Kiev. If they hadn't tried to take back their territory, he argued, and just negotiated with the rebels, everything would have been OK. Alexander Boroday, the Prime Minister of the self-proclaimed Donetsk People's Republic (DPR) in Eastern Ukraine, added to the U.S. sanctions list a day before the airliner was shot down, said, "If it really was a passenger airliner, we did not do it."[2]

Both responses sounded weak and uncoordinated, almost destructive to their own positions. The Russian news service repeated the preposterous story that the Ukrainians had tried to shoot down Vladimir Putin's airplane and missed, hitting the airliner. They stuck to it even as the videos streamed out, making it look impossible. This is a problem of credibility.

CNN's case was much stronger, supported by actual events as recorded by locals and reporters. At times it was a little rough, particularly the discussions of body parts rotting in the sun because the nationalists barred their recovery. At times it was speculative, especially the inferences that the local nationalists might not have been capable of firing this kind of missile, and that Russian military may have manned the vehicle.

The Russians tried another story that the plane was shot down on purpose and contained only people who were already dead. This variation suggested the government in Kiev was behind the downing of the plane. This story got no traction because even the Russians couldn't believe it.

Tom Forman, one of CNN's finest, reminded viewers that the evidence was only circumstantial. Because so much of news today is "pool" reporting, where one station picks up the stories of other news broadcasts, the same reporters are seen on every major news outlet. The pictures are the same and repeated often. They come out quickly. They leave little doubt of the facts and reach millions of viewers worldwide.

This kind of scorching publicity was not the kind Vladimir Putin was looking for. He got calls from several world leaders wanting to know what he was going to do to get the nationalists under control and get the investigation of the crash site moving along. A Russian press conference on the recent events took viewers to a command center that looked strikingly new. It was clean, white-walled, and had only three men and two women in it. There was none of the usual clutter found in a military unit doing real work. The graphics were crude and very unlike a genuine command center. There were large video screens with pictures of aircraft and commentary on how the events might have occurred. Speculations that the Ukrainians were trying to shoot down the plane of Vladimir Putin and missed, or that a Ukrainian plane shot down the airliner by mistake, were incredible. The only people who would believe this kind of story are the ones who believe similar stories we see in magazines on our grocery store shelves.

What the Russians were doing was "sheep dipping." Historically, the term *sheep dipping* was used by intelligence services to define actions taken to create a new background while cleaning up an old one (which might apply here too); it was also used in educational circles to describe throwing out alternative approaches to teaching an idea. The Russians were sharing their thoughts out loud, throwing alternative theories around the globe to see which ones might stick in the minds of their viewers. Truth was not a factor in the decision to present alternative theories, some even conflicting with each other. There seemed to be no visual evidence to support what they said.

There are two things that work well in information war—currency and credibility. CNN had them both; the Russians had neither. The U.S. and other foreign governments were behind both of them—too slow to keep up. If there is a lesson in this, it should be that speed and accuracy count, and may even

4. The War Peeks Out

make for currency and credibility. This is not the same as telling the truth, because speed and accuracy are relative terms in this context. The Russians were getting stories out fast, but they didn't appear to be credible. The Ukrainian government was getting its version of the truth out a tiny bit slower, but it sounded much better. The Ukrainian version of the truth is easier to believe.

On 22 July, CNN announced that the nationalists were going to turn over the black boxes to the investigative team. The bodies were being collected and put in the customary body bags. The inspectors were able to look at some of the aircraft components, some of which had been moved or cut into pieces, "to help with the removal of bodies," the nationalists said. Some of the rougher-looking, less disciplined forces were replaced with soldiers having better fitness and bearing. When they spoke, the nationalists listened.

On the same day the airliner was shot down, the Israelis went into Gaza. Embedded reporters went with them, but they didn't stay with them for long. The troops were nervous, and it showed. A reporter wore a night-vision device, but pointed out that the Israeli armed forces did not have them in great numbers. This would be a military secret in some circles, but it was obvious when the troops were shown assembling without them. To compensate, the forces fired red-orange flares several hundred meters into the air over the cityscape. The visibility was as good as a night-time soccer match in any major city, except the lighting was orange—a strange color. We heard gunfire, far away somewhere, as the troops moved forward. We saw only a few embedded reporters in combat units after that.

In the buildup to this part of the war, the most visible signs of conflict were white contrails of rockets fired from Gaza, and the Iron Dome low-altitude, anti-missile system flashing up to intercept them. The United States had helped develop that anti-missile system and the Prime Minister expressed his gratitude for their success and U.S. help. They hit some, and missed others, but it was impossible to tell the frequency of either from crossing contrails. In an ominous sign, the Israelis shot down a Hamas armed drone with a U.S.-made Patriot missile. Drones, even those of Hamas, are now carrying missiles.

We saw a graphical depiction of events—the SA-11 shooting down an airplane and the Israeli missile striking the armed drone. These were not real news—not actual events—but simulations of what might have happened. They were mixed in with real footage of events to help fill in the gaps of what we couldn't see. Like some video games, the graphics were almost as good as the real thing. The casual onlooker had to look closely.

The Palestinians had their news moments, too, firing rockets into Israeli cities from their side of the border. They showed moving videos of civilian causalities, their major weapon in this war. Children with blood on their clothing were crying as they were taken to safety. This became a recurring theme. One man scoffed at a leaflet dropped in a white cloud by an Israeli aircraft. The papers said civilians should evacuate the area for their own safety, but the Israeli infrared pictures showed a mixture of men, women, and children on the roof of buildings where Hamas kept some of it leaders. The cyberwar is on.

The Israelis said these attacks were targeting tunnels built by the Palestinians. A night-vision camera showed seven armed men climbing out of a hole in the ground, followed by a flash as an explosion lit up the camera lens. The seven were gone, more or less, but the video cut out just in time for us to miss the details of their destruction.

Al Jazeera America[3] showed efforts by Hamas to try to get locals to stay. It also showed the lengths the Israelis went to avoid civilian casualties. They called people in the building that was going to be blown up. They dropped a noise-making bomb on the roof, and waited. (This is called "door-knocking," to get the attention of anyone still around.) They then dropped a real bomb and the building became rubble. The Israelis knew the real weapon the Palestinians had was their own dead—a morbid, but truthful, assessment. Public opinion, translated into political action, attempted to get the Israeli advance to halt so casualties would stop. Israel couldn't help but hear what was being said.

We saw the fear in Palestinian eyes as parents bundled their families into cars and drove off. The occasional male said he would support Hamas against the invasion. He said he was staying (perhaps not the best decision).

Hamas and Israel traded images. Hamas showed a sketch of a soldier being held in the hand of a cartoonish fighter, saying one had been captured. They released a YouTube video of a tank gunner hit by machine gun fire. The Israelis countered with a picture of the UK parliament building with missiles falling on it. The caption reads, "What would you do?"[4] Days later, Shimon Perez, the former Israeli leader, used those exact words to defend the need for troops still in Gaza.

Near the end of this conflict, a missile from Gaza landed a mile from the runway of Ben Gurion Airport in Tel Aviv; the airlines subsequently reacted to the United States cancelling flights into the airport. Hamas celebrated a "great victory," and Israel complained about being isolated for no good reason. The cancellations ended the next day. The airport was apparently safer than the day before.

In the post-battle polls, the party in power in Israel, Mr. Netanyahu's Likud faction, got an uptick in favor of their actions, but 58 percent of Israeli Jews believed it was a mistake for the government to accept an open-ended cease-fire with Hamas. Sixty-one percent didn't think Netanyahu achieved his goal of prolonged quiet.[5] It looks here like the will of the people is to fight and not give in to new peace initiatives. We know what is going to happen next; it is harder to say when.

4. The War Peeks Out

The two sides of the larger war thus came together in one day. These two politically and geographically separated events are part of the same war. Our perception of these events may be that they are representations of regional conflicts that affect only a few people, but CNN and Al Jazeera America bring that to our home. We are only seeing the lowest levels of war.

The layers of war are deceptive. At the highest level, there are political differences between authoritarian governments, especially Russia and China, and the free world (the United States, Great Britain, Canada, Australia, and New Zealand, who work together on law enforcement, intelligence sharing, policy, and economics). The latter are the so-called "five eyes" countries, which disagree on various issues but work most of them out, or agree to disagree with no harm to any of the participants. These are countries that are not at war with anyone, but have fought many battles.

There is also a coalition in NATO where information is shared between the 28 countries. Of interest since the Ukraine situation have been new NATO members Albania, Bulgaria, Croatia, the Czech Republic, Estonia, Hungary, Latvia, Lithuania, Poland, Rumania, Slovakia, and Slovenia. NATO is a mutual defense and political organization, formed before those countries became members, in order to keep Russian expansion from going further than the boundaries of the old Soviet Union. Whether NATO could be effective at constraining the Russians is still a good question.

The two regional conflicts discussed earlier will be old news in a month or so. We forget about the events quickly enough, but we still apply resources to finding out what the Russians and Chinese will try to do to us tomorrow.

For all of my government career, and up to the present day, we were at war with Russia and China. I knew we were at war because the vast majority of our military and intelligence efforts were directed at Russia and China. My bosses in the military and civilian sides of government told me to watch what they were up to.

That hasn't changed just because we had "resets" with the Russians and made China one of our top trading partners. This is the main difference between the political administrations, the diplomatic corps, the militaries, and the intelligence communities of countries. The militaries and intelligence functions know where war will be, if there is one, because they are the ones who have to prepare and fight. Even when these countries are having good relations with the rest of the world, other governmental bodies are still trying to figure out what they are really doing. They have good reason.

Most of China's and Russia's allies cause trouble in the world, and we plan operations to counter them or disrupt their ability to make mischief. North Korea can always be counted on to do something dramatic when nobody else will. The North Koreans make a lot of trouble in various ways, but it is difficult to believe that China doesn't have a hand in what they do.

China is the only friend North Korea has in the world—other than Dennis Rodman.

The attention given to Russia and China is based on something called a "threat assessment," the latest annual one having been made public by the Director of National Intelligence (DNI).[6] These reports are remarkably similar for any country, though almost devoid of substance. In international sharing, each country publishes a threat assessment and other internal documents that are classified national security information and not shared with the public. The threat is clear to those who are allowed to see the documents, but they are top secret, so readership is limited.

What the public sees are general statements of large subjects that fall into three categories: terrorism, cyber, and nuclear proliferation, all embedded with regional issues of concern. Al Qaeda figures prominently in terrorism concerns, along with regional subjects. The United States thinks the Lebanese Hezbollah is a threat to "U.S. allies in the region," but Israel hardly ever mentions it. Great Britain, for example, sees its terrorism threat linked to roots in Ireland, as well as Al Qaeda, but the international threat is similar for each country that prepares an assessment.[7] Nuclear concerns are focused on Iran and its development of a bomb.

Threats are relative, so there is some sharing of threat information of common interest. The NATO-Russia Council even shares some information on issues related to drugs, missile defense, and counter-proliferation. China has an interest in the security of Afghanistan after the allies leave and does not want to see the Taliban become a threat. Hamid Karzai has been to China five times since 2013, and U.S. diplomats have had regular discussions about maintaining stability in the region, a concern shared by both the United States and China.[8]

The DNI report is done by categories of threat, and it is just for the United States. These are things the DNI thinks are the most dangerous to our country's well-being, so there are a variety of them, even some we don't think about very often. It is the only threat assessment that calls out Russia and China as specific threats. We know where to concentrate our efforts.

There are two types of threat—global and regional. Cyberwar, stemming intelligence efforts directed at the United States, weapons of mass destruction, space (as in outer space) operations, transnational organized crime, economic and resource competition, world health, and mass killings characterized as atrocities are the global threats. These are big things to think about, and any country would find it difficult to put large sums of money into attacking all of them at one time. There are also several regional threats, especially where people are fighting right now. The conflicts in the Ukraine and between Israel and its neighbors are regional, and although they are hot spots in 2014, they may be forgotten in a year. As ISIL grows, we quickly forget about these other regional battles. Global threats seem to not go away quite as easily, and have greater consequences if something isn't done about them.

The first type of global threat is cyberwar and the activities of people who use cyberwar's capabilities to undermine the stability of a government or part of an economy. The Worldwide Threat Assessment uses language most people don't use in each other's company to describe an element of threat, so we have to be patient to slog through the terms they use to describe events:

> Russia and China continue to hold views substantially divergent from the United States on the meaning and intent of international cyber security. These divergences center mostly on the nature of state sovereignty in the global information environment states' rights to control the dissemination of content online, which have long forestalled major agreements. (Page 2)

When we get to the real issues in this "substantially divergent" view of theirs, we find the DNI concern to be exploitation and disruption of infrastructure components, similar to those suffered by South Korea in March 2013, when tens of thousands of commercial business and media network computers were damaged. This attack was attributed to North Korea by most of the press reports and was said to have cost over \$550M to repair.[9] That would make it one of the largest and most damaging ever known. The report also mentioned the Iranian attacks on the U.S. banking infrastructure (Operation Ababil), among others.

We have enough evidence to be positive that somebody is looking into the U.S. oil and gas infrastructure and the electrical grid. They are using trojanized software to get access to key systems and people, doing what the military calls intelligence preparation of the battlefield.

If my army is going to fight a war with China, I need to know what that war will be like and how my forces will fight and win. I want to know everything about the military systems and strategies used by the Chinese. There are troops, planes, tanks, artillery pieces, transports, ships, missiles, drones, their communications, and a host of other things that I want to know more about. When the Chinese steal in such great quantities and frequencies as they are doing now, I should be concerned that their military is preparing for war.

The military that is going to face these weapons wants to know more than simply what they look like, and feels justified in expanding its role to find out every detail about every weapon and capability. Looking at pictures from a satellite will only tell the army so much.

I was once sitting in a meeting with some of the leaders in Ballistic Missile Defense while they discussed how they thought a missile made in another country might perform. There was a lot of back and forth, until my boss, Dr. Tom Ward, asked a good question: "Why don't we get one and find out?" I was thinking, "And why would these guys give us one to look at?"

In fact, what my boss knew that the rest of us didn't was that arms merchants don't have the same affinity for a government that a government has for itself. The line that separates economic espionage and military intelligence is wide and grainy because arms dealers have their own political and economic agendas not set by the governments they serve. We did eventually get a missile and test it, just to see how good it was. It wasn't that hard to obtain.

The point was that we didn't have to steal anything; we could just buy it. The world's various militaries get weapons from other countries to test and evaluate, and it keeps everyone informed about what they face in battle. At the same time, it gives arms merchants a market for new products.

It is reasonable to believe that military reconnaissance of computer networks is necessary to get a good picture of the battlefield, if parts of the battle are going to be fought there. We want to know what computers do and how we can disrupt them, spoof them, or use them to allow our army to win. Almost all of this reconnaissance is cyberwar, but a part of cyberwar that no government likes to talk about.

At the national level in the United States there are certain networks that provide secure voice and data communications for those in leadership positions, like the President and the Joint Chiefs of Staff. Operating and defending those networks is a military function that is also part of cyberwar.

At the same time, knowing how to disrupt those secure voice and data communications channels in a potential enemy country falls into the same category. We don't want to make disrupting our communications channels easy to do, so we have policies to protect our most sensitive networks. The United States has been concerned about IBM selling its X-86 line of servers, used extensively in some very sensitive networks, to China.[10] We know, based on China's past interest in hacking computers, that having them manufacture systems used in sensitive networks is a bad idea. We should know better, but seem to put "business interests" above national security in some situations. These are places where business interests have won out over national security concerns, with very little debate.

The vast majority of networks between businesses, financial services, and other critical infrastructure services, and even within the government, are not military networks, or national networks doing command and control. The military has no role in security for them. Yet they are also part of a cyberwar.

So, if my military is scouting around in another country's computer network, and goes into university research facilities, Internet companies, libraries, medical facilities, public utilities, businesses and government offices, are we OK with that? My army says, "We have to know everything we can,"

4. The War Peeks Out

as its justification. "War will be everywhere." If my army were protecting you, you would surely agree that I should know everything about our potential enemy in order to increase my odds of keeping you safe. You expect a swift and decisive victory. We should think about that more.

Intelligence services have the role of collecting and analyzing data about other countries. They have operations that require them to engage in the same kinds of activities as the military, and those operations can be undone if another government starts looking more closely at their networks. Too many operations against the same targets will lead to the operations being discovered. Intelligence operations have to be limited and controlled. In the United States that control lies with the President.

Intelligence operations cost money. Every country has a budget, and in spite of popular misconceptions, they are limited in the amount of spying that can be done. Those limits are prioritized in national collection strategies. No government wants to collect the same information more than once; it leads to additional cost and confusion. To manage costs and limit redundant operations, spying is usually left to the intelligence community, while intelligence collection to analyze military capabilities is left to the military.

In days past the Russians had the KGB and the GRU. The GRU did the intelligence collection regarding military capabilities and the KGB did everything else. We can't say there won't be overlap, but since they are both stealing from us, we ignore the differences.

If we take into account that the will of the people is important to winning a war, it becomes a little harder to put stakes in the ground and say who is responsible for that aspect of war. It depends to a great extent on the kind of government a country has. A ready example comes to mind.

What most Western militaries lack is something few democracies have—a censorship bureau, required to enforce certain reporting and commentating standards on the general population. The United States, which is often thought of as a leader in free speech, actually considered having one for some of its government employees who have access to classified information, but dropped the idea when it met with too much criticism.[11] Censorship is a kind of forbidden word in democracies and, in the days of Skype, Facetime, and text messages, extremely difficult to accomplish. The common euphemism for censorship is "operations security," which is generally applied to two areas—intelligence and military operations—but not to a whole country's population.

A country that controls information flow and infrastructure can control some of the content, especially the part that influences the will of the people in a fight. Controlling the will of a population means controlling communications between individuals and influencing them, either directly or through mass media. We might not like the fact that a country does this, but we have no control over it. Whether militaries have a role in this process depends more on their internal organization than on their capabilities. Every country has to manage the message to the populace if they are going to win this kind of war.

Democracies generally do not publicly allow their militaries to be involved in managing messages to the public, nor those to the populations of other countries they are trying to influence. Most democracies are not happy about admitting they do this. Yet, in every one, the central government defines the level and type of influence it wants to have, and economic and political parts of government carry it out. In the name of national security, every country does things its citizens might object to, if they knew what was being done.

One of the first claims made by Edward Snowden was that the United States was collecting messages from Chinese cell phones and getting Blackberry text messages belonging to G20 members at their 2009 summit.[12] The referenced article claimed China was so concerned about this issue that they began replacing network components with Chinese-made items, an irony hard to miss. China's Huawei is accused of doing similar kinds of things as a part of their intelligence services.

Both Russia and China have similar views on the control of the Internet to further their economic aims and preserve their control over their own political power, including their military capabilities. Both are considered to be the most active at "penetrating the U.S. decision making [sic] apparatus, defense industrial base, and research establishments" (Clapper, page 3).

The effort to get into the decision-making apparatus might attract our attention. It is the first time it has been mentioned in the context of threats to the United States. This may be a reference to our politics, our business leadership, or both, but in this version of the DNI report, we don't get much detail. If the threat assessment is too specific, it leaves no room for Congress to disregard it. Having foreign governments involved in our decision making at the national level is something we cannot ignore.

We should remember that a large part of what is popularly called cyberwar is really transnational crime, defined as war by places like Uppsala University, where conflicts produce casualties in sufficient numbers. Organized crime is using (and sometimes even inventing) new ways to attack systems, and those methods help push cyberwar's technology. Transnational crime is more than pornography, prostitution, and drugs. It can also be the development of new, artificial drugs that don't require poppies to manufacture. Synthetic drugs can make for better health or better highs, depending on how they are made and distributed. The dark side of networks is represented by "dark nets" like Silk Road—non-attributable networks used to hide illegal transactions from law enforcement.

Silk Road used to be hidden but, since its owner was indicted in February

2014, now finds it difficult to stay out of the spotlight. Silk Road was run by Ross William Ulbricht, more widely known as Dread Pirate Roberts. Ulbricht was first indicted by his pseudonym in a murder-for-hire case. The Pirate had attempted to contract with an FBI agent, a Silk Road user, for the dispatch of a man who threatened to disclose the names of thousands of network users. The FBI contacted Icelandic authorities in May 2014 to help with leads on servers being operated in that country. That investigation led to servers in Pennsylvania.[13] Silk Road was designed as a secure network that could only be accessed with a Tor anonymous browser, and then only by people who knew where the network was. The *New Yorker* magazine used the same principle when it set up an anonymous network for confidential sources.[14] Silk Road accepted only Bitcoin for its transactions and took in tens of millions of dollars. Silk Road also sold hacking tools, though its major source of revenue was drugs.

When the first Silk Road operation was shut down, it quickly grew back to the size it was when Ulbricht was arrested, but it eventually lost market share to two other sites, Agora and Pandora. The use of the dark side of the Internet isn't going to stop because of the arrest of a few people.

On a grander scale, the revenue from conducting criminal operations helps fund regional conflicts or groups that can't get funding from other sources. Some of that money goes to terrorists, either through their participation in the enterprise or through pay-offs for protection. Competition for drug money creates gangs in Latin America that push citizens to leave and go north to Mexico, the United States, and Canada. The gangs create a threat to U.S. border security that is blamed on the Obama Administration. In Afghanistan, it keeps money flowing to the Taliban. In Russia, it feeds the economy through cybercrime, sucks the life out of its own citizens on drugs, and corrupts public officials.

Cyber is related to another international threat—weapons of mass destruction (nuclear, chemical, and biological). Science offers salvation to many of the world's people, but it also has a dark side. Science makes it easier to transport nuclear weapon designs from one place to another in easily hidden, or hidden in plain sight, electronic devices. It likewise makes it possible to create a plastic gun on a 3D printer. Science spreads genetic transformations to places in the world where they can be misused in ways we haven't thought of yet. It brought us into contact with Stuxnet, the Energetic Bear, and other forms of infrastructure attack that, as we have seen since, are not easily controlled when they get out of the lab. What it also shows is the inability of our intelligence services to neatly categorize complex, interrelated issues.

Outer space is the one threat area that is predominately still in military hands. Both the Russians and the Chinese have anti-satellite weapons and jamming capability, but most of that is the responsibility of their armed forces. Commercial companies do work in space, delivering cargo and putting up satellites, but jamming them or shooting them out of orbit is not something they do to one another. In a kinetic war it is possible that satellites we use for phones and global positioning will not be around. This won't affect the military as much as the attacker would want, but it will affect the average Google user who relies on global positioning systems.

Competition for food, water, and energy is intense in some countries (like China) that rely on the external world for most of their supplies. With Europe on one side, and China on the other, Russia stands to benefit from shortages of energy and food. When a country gets 30 percent of its energy from Russia, it has to think twice before being too critical of Russian politics.

China tends to link its political support to energy contracts in places like Angola, Nigeria, Iran (where China suffered setbacks in development of the Azadegan field near Iraq[15]) and so many others that it is impossible to cite them all. China has relations with almost anyone that sells energy, including U.S. companies and universities (for research on making its energy go further). China and Russia signed a $400B contract for gas in May 2014, in the midst of the Ukraine fighting, which is costing Russia customers who are looking for alternatives.[16] The Russians have built new pipelines across Siberia to reach Chinese markets, but they are asking China to pay more for infrastructure development (a point not settled in the original agreement).

Food competition influences fighting in the sub–Saharan parts of Africa where Cheick Aoussa lives. Deserts grow and grasslands shrink, putting the squeeze on farmers and cattle ranchers. This causes friction that generates the kind of heat that will pay him well just to take sides. We can imagine that he will take advantage of that situation.

In the South China Sea, where Viet Nam, the Philippines, Japan and China look for energy, the fishing is not as good as it used to be, forcing boats to travel further into territory that is claimed by multiple countries. They clash, making trouble for all of them. The Russians and Chinese hold joint military exercises to show Japan that they are cooperating. The Chinese get their first aircraft carrier. Both raise their defense budgets to gain ground on the countries that compete with them.

Diseases that we once thought were eradicated are now back. We have smallpox, drug-resistant strains of old diseases, polio, measles, and the largest outbreak of Ebola the world has ever seen. Ebola has killed over three thousand people, and the World Health Organization has yet to stop the spread in Liberia (which has enough problems), New Guinea, and Sierra Leone.[17] The technical ability to perpetuate disease, or manufacture new ones, is never very far away. However, diseases are not yet very good weapons, since anyone who starts that ball rolling has to be able to survive the consequences in their own region or country.

If it seems like we are living with more trouble than we used to have, that's because we are. Adding nuclear proliferation to the world's other threats makes the nuclear threat seem smaller than it really is. A country like Israel, with six million residents, sees it as a problem of survival, not a difficult political problem that needs to be worked out. Not very many countries want Iran to have a bomb of its own, but the Iranians keep marching in that direction. Stuxnet delayed them,[18] as did having some of their nuclear scientists assassinated, but they have patience.

The greatest threats to the United States are traceable to Russia and China, but we have to be careful about how that statement is inflected. We could infer that Russia and China are one—that they think alike or are engaged in some kind of gigantic conspiracy. Only a few people, on the edges of society, believe that is true. They are not conspiring, but they are not getting in each other's way either. They find reasons to cooperate publicly and minimize their disagreements.

Richard Weitz, a senior fellow at the Hudson Institute, points out that Russia and China are symbiotic.[19] They have similar views on Asia-Pacific security, Iran's nuclear program, and Syria's civil war. Both countries have a bad feeling about Islamic fundamentalists in the border areas of their countries, and they are a lot closer to these areas than most of the rest of our world. They also have shared concerns about Kazakhstan, Kyrgyzstan, Tajikistan, Turkmenistan and Uzbekistan, which most of us cannot point to on a map. Neither country likes the U.S. attempt to make every country democratic, even when it isn't possible, nor our occasional reach into the internal affairs of both countries—and a few of their allies. They believe Russia's weakness after the revolution allowed the expansion of NATO at Russia's expense.

Both China and Russia say the relations between the two of them have never been better, then smile and shake hands. They even conducted joint military operations exercises for the second year in a row. These exercises are not particularly large by exercise standards, but they are symbolic by their location and include special forces. They are practicing live fire (i.e., with real bullets) and coordination of forces in the Sea of Japan, where the Japanese have increased their presence.[20] There are a few islands in dispute, though it is hard to call some of them disputed since they are hardly above water at certain times of the year. The Russian-language newspaper *Kommersant* reports that China and Russia are about to start joint cyber operations, but it says little about what might be done or where.

Having some things in common doesn't always mean that Russia and China cooperate in the ways that allies do. During the Cold War we assumed the East Germans, Czechs, Albanians, Latvians, and Russians were like a Band of Brothers, doing everything together and sharing like brothers do. When the Cold War ended and intelligence reports were dug out of vaults and crevices, it turns out that they didn't cooperate nearly as much as we thought, and some of them were working, now and again, against their brother's best interests. We can shrug that off, saying that even the best of friends don't always share. That would be true here, too. China and Russia have a long, common border pretty heavily defended on both sides. They compete for energy in some of the same markets, and Russia backs Viet Nam in its claims of territory the Chinese would like to have. There is oil in that water, and the Chinese and Vietnamese would both like to have it. The Russians also dislike some of China's incursions into their Far East, fearing it will become part of China one day.[21]

The fear of incursion is rooted in another incident that took place in 1969 on the Ussuri River. The U.S. State Department, in a declassified memo, described the events as a "clash" between Russia's 25 Divisions in the area and Chinese troops remaining from the Korean War. Both Russia and China described it as an attack by the other, but Russia used the term "ambush" in its description.[22] China, however, claimed these skirmishes had been going on in different parts of the river for over two years. They had demonstrators outside the Soviet embassy in Beijing with signs that read "Hang Kosygin" and "Fry Brezhnev." These are not the kinds of things friends say. They claimed an 1860 treaty was not fair to China, indicating that the Chinese carried a grudge for a long time.

When the two sides finally got together to settle the dispute, it was for an odd reason. The Chinese saw the United States in Viet Nam, and were unsure of its intensions. They were pretty sure they knew the Russians better. In 1999, the Public Broadcasting view of events provided an interesting perspective: Beijing had decided to take the advice of Mao to "settle with the enemy far away, in order to fight the enemy at the gate."[23] The Chinese chose to make Richard M. Nixon a friend and Henry Kissinger a world traveler.

While we may view this as a historical quirk, Weitz says, "The Russian military is also undertaking its own Asian pivot. Although Russian rhetoric is directed against NATO and the United States, Russia's newest weapons now typically flow to eastern Russia."[24]

The Russians and Chinese are increasing their defense spending at a time when the United States is cutting its by huge amounts. These numbers are deceptive.

Defense is about 4 percent of the U.S. federal budget, and falling precipitously, but it doesn't include the budget of Homeland Security. OMB includes in its analyses a category called "national defense," which was 19 percent of the budget in 2013.[25] This comes closer to a description of the kinds of things done to protect our nation. Anyway the pie charts are drawn in a budget measured in trillions, it is a lot of money.

But what we don't usually realize in discussions of the budget is how the United States compares to other countries. In Eisenhower's time, very few countries could match us on budgeting for national defense. Today, very few of the ones that might will tell the truth about what they spend. NATO is financed largely by the United States; in 2013, it was paying for 73 percent. In another example, China says it spends $119B on its army, but China's army is much more than a federal activity like it is in the United States. Its total internal security budget is greater than the U.S. spending on defense.[26]

China's government, sometimes its military, still operates its own businesses, some of them making billions of dollars a year. The Chinese claim to have corrected the military operation of businesses, but still have trouble making it happen. Now they call making money from these relationships "corruption." They don't count that in the military budget, or, some analysts say, their expenses for big-ticket items like their only aircraft carrier. So, they really only tell us some of their expenses. For our part, we still have a "black budget" that we don't tell them about. Our budget used to be top secret, and we pretend nobody knows much about it. In fact, it has not been a secret since Edward Snowden gave it to the *Washington Post*, which published it.

We should be even more skeptical about Iran, which indicated that it wanted to increase its defense budget 127 percent from about $15B,[27] in order to counter a possible attack by Israel. Israel spends almost $58B, which includes $3B given by the United States for that purpose.[28] Russia says it spends $65B but plans to increase that number to $105B in 2016 to focus on upgrading its nuclear weapons.[29] Britain is around $54B, and France is not far behind. It isn't the individual numbers that make mercenaries out of the defense industries, though; it is the aggregate. The arms business is huge, especially in the United States, where the bulk of the world's arms trade takes place.[30]

President Eisenhower was not the only world leader to be struck by the power of military-industrial ability to influence world events, but he may have been wrong about concerns that the complex might usurp our democratic principles. These days, money that comes to business is often a national security matter that makes for a significant part of modern war. Our latest example is the National Security Agency and cyberwar.

On a global basis, we fight wars that didn't exist in von Clausewitz's time because he didn't have the technology. The networking of computers changed that, but it didn't change things equally for everyone. In the early 1980s, when the doctrine for information war was being formed, there weren't many computers, and those that existed didn't represent very many people compared to today.

Now we fight over ideas, not territory. The new mercenaries mask the real people at war, and the kind of war they are fighting. The business of war looks at who benefits from the fighting and how we find our enemies. The whole understanding of von Clausewitz's definition lies in knowing the enemy—so we know the one who bends to the will of the victor.

We still have a tangential idea that war is the failure of diplomacy. Arthur Schlesinger, Jr., in his evaluation of two books about famous diplomats, focused on the demise of the diplomat, who is now chained to electronic devices and has more help than one would ever need. But the realization has finally come that the failure of diplomacy and sending people to die for that failure is now more of a public phenomenon. More people have input into those decisions; more people feel they have a right to engage. The diplomat is always behind, and never home.[31]

We can say the professional soldier is in the same boat as the diplomat. We have to wonder if the soldier of today understands "chain of command" outside the context of the string of CC's on the email line above the subject heading. Some of those are political appointees, and Congress members and their staff, in various government agencies who never served in the military and have no business being involved in military decisions. Both the military and diplomats would be more inclined to see war as a failure of communication when there are more than enough means of communicating. What we may be missing is the idea that war is only a failure of one idea to win out over another.

5

CYBERWAR

Cyberwar is about information control, those who have it, and those who don't. It is the major component of a new style of war that is strategic, thoughtful, and long term. Governments are the only entities that fight this kind of war. They control information because modern war integrates the technology of managing information, business interests, and national policy with the need to dominate the will of another country in order to win.

This kind of war attempts to control both the media used to manage information and the content of messages those media pump out. That means managing the press, social media, and thought leadership in policy areas. Nobody will say this is easy, or that it can be done completely, but management of both internal and external sources of information is essential. These components of cyberwar are far from the purview of most democratic countries' militaries. The majority of their leaders won't even admit they do it.

What makes the governments of Russia and China enemies to the free world is a battle for the will of people—not just the people inside their own countries, but everyone who has any expression of views that differ from their own. They want to control the content of speech and text, even if it means undermining other countries' institutions to make that happen. We have a right to be offended by such intentions.

As part of their state policies, China and Russia steal on a scale never before experienced between nations. That is a separate matter from internal information control. Stealing economic information and using it to expand their domestic economies is short-cutting the development time for new products and services; that is largely economic warfare. If the purpose is to give their leaders more accurate information to make decisions, it is spying, though we might have trouble making the fine distinctions required. Both Russia and China provide a protected environment to those stealing outside their individual economy or collecting business intelligence. The same type of protection is required for both tasks.

There is an inferential leap required to see war in what these countries do. The fundamental rule of war is summed in Carl von Clausewitz's words: *War therefore is an act of violence intended to compel our opponent to fulfill our will.*

In von Clausewitz's time, nobody had heard of communications like we have today. The winners in his day had control of the means of communication and the ability to directly influence the content. They controlled the content by controlling territory and distribution of written ideas. That is harder to do now because there are more sources, they are all electronic, and they cross national borders. They go directly to the audience. Von Clausewitz's forces would have been frustrated at trying to control something so complex. In his day, control of the message was easier.

The Internet has given us a way of monitoring what the public says they want, regardless of whether they willingly supply that information. That is a powerful tool, no matter how it is used. Once in a while public reaction to something will be negative, and governments determine what the reaction should be. If the person expressing a view is put in jail for saying what he says, we can be sure other people will be hesitant to make similar statements, knowing what could happen. Even in the most democratic countries some will speak against the government, and their comments will be used as justification for isolation, criticism, or scrutiny. This may make others of a like mind think twice about posting the same kind of comments.

Although most of the world's people have access to mass media, only about half have access to the Internet. Those numbers have risen from a third to a half since 2000, because our Internet growth has exceeded the rate of population expansion. As more people are included in the World Wide Web, there are opportunities for cyberwar that didn't exist before. Looking at those numbers another way, we had less opportunity to use cyber techniques to win, or manage, the will of a population in 2000 than is available now. Most governments recognize that.

Governments have quite a few reasons for behaving as if their citizens need watching. Most of us don't care about their altruistic reasons for these intrusions, and surveys of Internet users confirm the reluctance of users to accept them, even within the most rigidly controlled countries.[1] Governments don't seem to mind that we don't agree with their position. In the business of cyberwar, it is governments who fight, and they do so without the consent of their citizens.

The medium used by governments are telecommunications services: AT&T, NTT, Verizon, Deutsche Telecom, Telefónica, China Mobile, Vodafone, France Telecom, and América Móvil being the biggest. Internet Service Providers buy circuits from telecommunications services. Together they deliver the Internet, and its converging services, to users. Each one has a tenuous relationship with the military and intelligence agencies in countries they serve. They get along with these agencies because they have to, not because there is a good business reason to do so.

There are two aspects to what governments want from these service providers: monitoring for national security, and identifying and locating criminals and criminal activity in their country.

Although we don't think about it very often, every government has a right to monitor other governments' intentions and its own population. We should look on these as public safety, in several forms. We keep records of traffic offenses, gun purchases, immunizations, sexually transmitted diseases, fertilizer and amphetamine bulk purchases, births, deaths, and most of those things that can be collected from computers and cell phones. We are not afraid of the collection of information, but we are afraid of how that data might be used. When Russia and China offer the same reasons for collecting this kind of information, as they did regarding the registration of websites to "prevent the spread of child pornography," they apply monitoring more broadly than just nude children's pictures on a website. Conflicting government approaches to how to manage their populations, while also influencing their enemies, are the fundamental disagreement that puts us at war.

But governments also have limits. Facebook found out that we don't like it when institutions use a medium to manipulate our perceptions. We generally don't approve of it; yet this is part of everyday war in the Cyber Age. The Russians focused on a few people in the Ukraine, controlled their access to communications, supported those they favored and discredited ones they didn't. They controlled the press through intimidation. They used their economy to leverage perceptions on people who had experience with that before. They brought back the Nazis. They made up stories of dubious credibility to cover their mistakes. In the end, they got the Crimea, and are still working on getting more. In the process, they are solidifying their enemy's response through the transparent manipulation of information. Nobody likes that.

Vodaphone published a report of some of its dealings with governments and opened a few eyes in places that probably should have known what these companies deal with every day.[2] The company omitted Russia, China, and the United States from its report but found plenty of others to talk about that wanted control over networks within their borders. What the report demonstrates is the range of control over service providers. Some countries demanded the ability to see into the networks at the content level (e.g., they wanted to be able to examine the words that are used between people using the Internet or phone service).

All countries have rules about encryption, especially the strength of encryption, and the ability to monitor communications when required. Companies and individuals use encryption to keep other people from seeing their data. Countries can, and do, get access to encryption codes for communications passing through their borders. The carrier can't say no. In China's most recent change to its counter terror policies, it is requiring companies to do two things: (1) turn over any encryption used to protect information internally and (2) any source code for software which might be used in China. The former is consistent with the requirements of most other countries, the latter is not. In almost every case, monitoring in the name of law enforcement is the reason given for requiring this capability, but once it exists, controlling it is another matter. The Vodaphone report describes it this way:

> In most countries, governments have powers to order communications operators to allow the interception of customers' communications. This is known as "lawful interception" and was previously known as "wiretapping" from a past era when agents would connect their recording equipment to a suspect's telephone line. Lawful interception requires operators to implement capabilities in their networks to ensure they can deliver, in real time, the actual content of the communications (for example, what is being said in a phone call, or the text and attachments within an email) plus any associated data to the monitoring centre operated by an agency or authority.
>
> Lawful interception is one of the most intrusive forms of law enforcement assistance, and in a number of countries agencies and authorities must obtain a specific lawful interception warrant in order to demand assistance from an operator. In some countries and under specific circumstances, agencies and authorities may also invoke broader powers when seeking to intercept communications received from or sent to a destination outside the country in question. A number of governments have legal powers to order an operator to enable lawful interception of communications that leave or enter a country without targeting a specific individual or set of premises.

In the United States the same activity involves the Patriot Act, Section 215, which gives the FBI authority to monitor business records associated with terrorist activities. But there isn't a country in the world that doesn't require the cooperation of telecommunications companies that operate within their borders. They need to have access to criminal activities and operations, spying, or terrorism threatening their national security.

This is made confusing by the mixing of military and intelligence functions in most governments. In the United States, the National Security Agency is the best example. The NSA is run by a military officer and includes a number of military employees. The fact that the NSA is part of the intelligence community, as well as the military, is not that unusual. This was part of what motivated the U.S. government to limit the NSA's authority to collect information from domestic phone calls. The President restricted the use of the information, but not the gathering of it.

> Alongside the invocation of privacy and restraint, Obama gave his plainest endorsement yet of "bulk collection," a term he used more than once and

5. Cyberwar

authorized explicitly in Presidential Policy Directive 28. In a footnote, the directive defined the term to mean high-volume collection "without the use of discriminates."[3]

Inside the United States, the military cannot exercise any law enforcement functions except over its own employees, or where expressly authorized. The United States has taken traditional and legislative action to prohibit military enforcement of law, although there have been some occasions (like the case of Hurricane Katrina) when military action has been considered.[4]

There is no similar restriction on the intelligence community. Most democratic countries have created domestic intelligence collection agencies to avoid the kind of issues spawned by the NSA's collection of data on U.S. citizens. Australia, Canada, France, Germany, and the United Kingdom each have one.[5] When the NSA performs this function, it can use its membership in the intelligence community as status for the work, disregarding that it is also a military organization under military authority.

Vodaphone mentions no specific country rules, but offers this summary of what is done in the name of national security:

> In most countries, Vodafone maintains full operational control over the technical infrastructure used to enable lawful interception upon receipt of an agency or authority demand. However, in a small number of countries the law dictates that specific agencies and authorities must have direct access to an operator's network, bypassing any form of operational control over lawful interception on the part of the operator.[6]

Vodaphone makes it clear that where a country has its own link, the carrier cannot control what information is extracted from Internet, cell phone, or other traffic.

However, China and Russia are not content to control their own internal sources of information; they also want to control anything that comes in from the outside. They are going after the sources of news and reporters who write stories regardless of where the information originates. These are borderless activities that are part of cyberwar, and they usually begin with the press.

How they influence the press in other countries is hard or soft intimidation. This is a little different from operating BBC, Al Jazeera, or Voice of America (which all put out news that favors positions of governments, although they are allowed some independence to challenge a government's position). In authoritarian governments, the focus is on internal people first, with military forces frequently involved. Their priority is staying in power. They have censorship bureaus (or some equivalent), and they have rules on what subjects are forbidden. The Chinese have a formal structure for their press that lays out what subjects will be handled, in what way, using what language. They send instructions to all press outlets or any media that reaches the public, and it is sometimes more detailed than just a choice of text or video. For example, the Rolling Stones were told they should not perform the songs "Honky Tonk Women" or "Brown Sugar" during appearances in Shanghai.[7] China decides what is good for its population. Internet Service Providers have to sign agreements to support the efforts of the central government and employ censors to help do that. We can be sure somebody is looking at playlists of songs.

When someone doesn't accept the guidance for press outlets, the consequences are greater than we might expect as observers from outside China looking in. In July 2014, a man was arrested and charged with "publishing on-line rumors," exaggerating the number of people killed in an incident in Xinjiang, where locals armed with knives and axes attacked a police station, killing 37 people; the police shot and killed 59 of the attackers.[8] Most of us would see this as a big event, worthy of some international coverage, but we saw almost nothing about it.

The person arrested had put the numbers of those killed at 2,000–3,000, but, more to the point, he had gone around the Chinese Firewall and published his accounts on external websites. This is a sin in China, going beyond the mere reporting of numbers. A pro–Beijing imam in China's largest mosque was killed shortly after the incident. The police launched a campaign to have citizens turn in anyone who might have participated in the attack on the police station; 18 people surrendered themselves.

But the real purpose of censorship in China seems to be "to reduce the probability of collective action by clipping social ties whenever any collective movements are in evidence or expected" rather than to stifle criticism of the government.[9] The Great Firewall of China filters websites that the Chinese find offensive; it filters key words (which the Chinese get around by using analogies and satire); and it censors things posted on the websites it controls. This is done by an army of screeners who review content and effectively remove (in less than 24 hours) what they don't favor. It might be more of a problem to be "trending" in China than in some other country, but it seems the Chinese are looking for stability in their population, not necessarily complete agreement.

There is evidence that the Russians have a less subtle way of controlling information, which is represented in the way they control the press. In the election of Russia's parliament in 2011, several websites belonging to Russian journalists and news agencies were hit with denial-of-service attacks directed at sites critical of what they described as election fraud.[10] Fourteen sites (including those of the radio station Ekho Moskvy, Russian-language newspaper *Kommersant*, and an election watchdog, Golos) were hit in a persistent attack that moved with rehosting efforts by the affected outlets. It takes a sophisticated attack to repeatedly hit a moving target.

Legal action was taken against at least one of the pollsters for publishing results within five days of the election, which is prohibited by Russian law (a curious law that benefits those in power). Sites that reported on-street demonstrations related to the election fraud were also disabled for a day, until the elections were over. This is mild activity, however, compared to the handling of reporters.

A few Russian journalists have shown up dead every year since Putin took control of the country, including eight in 2002. Most were shot multiple times, just so there is no doubt that the shooter intended the result. Yevgeny Gerasimenko, who may have needed to suffer more, had a plastic bag pulled over his head, and his body was stuffed in a duffle bag where it would later be found. A local man was convicted of robbery and murder in that case, but he was not linked to any of the other incidents.

In December 2013 Putin dissolved the popular RIA Novosti, replacing it with a press arm more aligned with public relations than news.[11] The new conglomerate will be called Russia Segodnya, and it will focus on foreign audiences, with emphasis on Germany and France, where there is some sentiment favoring Russia's perception of events.[12] East Germany (or at least the remnants thereof) certainly fondly remembers the old Soviet empire of which it was once part. At the opening of the Arabic-language branch of Russia Segodnya, the spokesperson, in an unfortunate choice of words, said, "Our Motto is: We don't translate the news, we create it."[13] In the future, we might look to how much creativity they show.

There is no such thing as an independent press in Moscow. We don't want to interfere with the way a country deals with internal matters, but the Russians don't stop there. They have shown in the Ukraine how they succeed outside their own borders. Local press people are being imprisoned, kidnapped, tortured, and killed faster than the news media can keep up.[14] The Syrians and many of their neighbors do the same thing. That controls the press reporters, but it does little to get to those thought leaders and politicians who influence others. They are working on that task in different ways.

Evgeny Morozov has written a good bit from practical experience of dealing with the former Soviet states, focusing on the Internet as being something more than the maker of democracies. When these states started, they censored websites and banned certain ones altogether, but Morozov observed them doing more. They were starting to pay their own bloggers to reflect views they want expressed, especially pro-government views.

The new states received technology from Western governments that would monitor people and social trends in close to real time. They used the Internet for intimidation and attacks against publishers, nongovernmental organizations, and individual websites. At the same time, they learned from the U.S. and European governments that the Internet, particularly Facebook and Twitter, provided good ways to communicate with like-minded people. We recognize the same thing.

Soon after U.S. attacks on ISIL, Twitter started to shut down ISIL-related accounts. ISIL, not to be outdone, and showing some understanding of the need for social media, re-created and re-advertised the accounts, only to have them shut down again.[15] ISIL used a website in Poland called JustPaste.It. It was easy to use and run by one dedicated administrator, who suddenly found himself in the middle of an international terror operation (and not entirely sure of how it happened). British police helped him find out. He started by taking some of the more graphic videos, including the beheadings ISIL uses to incite fear, off his network. He worked all day on his little site to keep the violence away.[16]

Morozov uses the example of the Iranian presidential elections in 2009 to make his point clear.[17] The State Department asked Twitter to delay maintenance of its network until after the election so the Green Revolution could continue. The Iranian government aggregated photos of demonstrators from Twitter, asking other Iranians to identify them. They suppressed certain Internet sites while also getting the official government views published. Morozov rightly says that our belief in a neutral Internet is a trap.

When Hillary Clinton gave a speech to encourage use of the Internet in other countries, and Michelle Obama used it as the subject for her Chinese visit, they were saying that the Internet is a good thing. They ignored how those governments will use the information collected from it if their audience uses it for things the government does not approve of. They may have thought that every government is like their own, although they should know better. Some countries put a sieve on the content of the Internet and use it to find those who might not agree with the government's position. They aggressively enforce manipulation of what a person might say, especially where that involves influence over large groups.

The Chinese found a dissident, Michael Anti, because Facebook requires the use of a real name, not a pseudonym.[18] The Iranian government published a story during the Green Revolution that said Neda Sultan, whose bleeding body almost everyone remembers because it was seen by every news outlet in the world, was actually shot by one of her fellow protestors. The goals of the Green Revolution were never realized, but the uprising awakened the Iranian government to the power of social media and how to use it for their own purposes.[19] Not everyone sees the Internet as a friend.

A Transition to Cyberwar

Controlling information that appears in public is more difficult than it looks and considerably beyond the scope of cyberwar as it was first defined by the military. Cyberwar was originally part of the larger context of warfare between nations. The military view had divisions for "cyber warriors" attacking a country's infrastructure to undermine basic services, and envisioned using computers to reduce an enemy's ability to fight. This aspect of war went in all directions to leverage military capabilities as "force multipliers." It was almost chaotic.

But while cyberwar was being redefined, the media of war was changing. We were seeing convergence in the various media that made up the delivery mechanisms. We now get television programming, movies, correspondence, news, pictures, text messages, telephone and quite a few other things from a single feed that is wirelessly transmitted anywhere in our homes. We become part of a cloud, those networks of mysterious origin and ownership, where we are given capabilities that cost us nothing. Yes, it is too good to be true, and we have to wonder how such good fortune has come our way. At some point we might want to remind ourselves "if it seems like it is too good to be true…"

What the Russians and Chinese know is that content management of those electronic services is far more complicated than it appears, but worthwhile when done properly. Russia used control of TV and radio stations in Crimea, as well as the telephone network, the Internet, and public advertising, to create the idea that the region was welcoming the invaders.[20] Russia can filter what people say to each other about its actions. It attacks websites that share unfavorable views. It introduces new ways to explain events, spreading those stories with media sources controlled by the state. It creates ideas that are simple and graphical, and spreads them through a multimedia campaign. It works a little like advertising at that end, but a lot like authoritarian management at the beginning. Syria, one of Russia's best friends, uses a different approach, but one that is equally effective.

Internet activists got involved in Syria, and they were not kind to Bashar al–Assad's government. A group called Anonymous (a mercurial set of crusaders with changing membership depending on the issues being addressed) launched Operation Syria. Assad closed down his Internet connections and phone service in May 2013, and Anonymous decided to intervene.[21] Governments generally don't like this kind of behavior, but we might want to give Anonymous some credit for taking a stand. Its members are trying to influence the outcome of a war, while acting without government protection. As a consequence, their results are scattered, but their detractors are not. We have to ask ourselves if what Anonymous and other independent groups are doing is acceptable. The answer to that question depends on whether we believe the outcome justifies the actions being taken.

Anonymous stole records from the Syrian railways, the parliament, the patent office, and Syrian TV, and it published these stolen items for anyone to read. The group also tinkered with the websites of Syria's embassies in a few countries. Other groups joined in, collecting and releasing more information.

Perhaps the most interesting thing was a set of records on how the Syrian government was monitoring its own citizens. Telecomix, another Internet activist group, released records of a monitoring tool called Bluecoat, software made in the United States. The software, which is a product of a company of the same name, has turned up in Syria, Egypt, Kuwait, Qatar, Saudi Arabia, the UAE, Iran, and Sudan. It claims 15,000 government and corporate customers.[22] Records from that software were used to find out what, and how, monitoring of individuals was carried out in Syria.

What researchers found was that the Syrian government used Facebook to steal the passwords and accounts of users, and then monitored the users online as they exchanged messages. The data collected would then be filtered and persons of interest could be developed. Government agents got their Bluecoat software illegally through a Dubai company to look for patterns in what they monitored.[23]

Research into where Bluecoat might be used started with fingerprinting the software (that is, developing a set of criteria that produces a signature that can be searched for on the Internet). In addition to the countries already mentioned, Lebanon had the software signature on its networks, as did Russia, China, India, Turkey, Iraq, South Korea, Malaysia, Thailand, Kenya, and Indonesia.[24]

Bluecoat can do a good deal more than just monitor Facebook. According to its own advertising, it is an all-purpose security tool that can monitor and control web interactions, phones, chat sessions, and messaging using content filters. A business has a perfect right to monitor its own communications, which is why such software exists. A user of a business's communications assets has no expectation of privacy, since he is conducting business on a company-owned system. There is so much information floating around in most business networks that sorting through all of it requires this kind of tool. Tools like Bluecoat are also used by law enforcement and security entities looking for criminals, to identify and map criminal social networks. When a government uses the tool to monitor its citizens, it is inferring that it has the same rights over its citizens as the business has over its employees and law enforcement has over criminals. The citizens might not agree, but they can't do much about it.

The Chinese approach monitoring social networks in a different way—by making their own Facebook, Amazon, E-bay and Twitter networks to control access by their users. It simplifies monitoring by building it in, a strategy the Russians are starting to discover.

A Peek at the Technology of Cyberwar

The technology of cyberwar falls into three categories: monitoring, exploitation, and attack. Monitoring and censoring large populations is no small job, as the Russian example shows. The World Bank lists Russia's population at 143.5 million; of those, Russia's favorite news channel says 55 percent use the Internet (though 14 percent use the Internet just a few times a month).[25] So, if the Russian government wants to monitor its own population, as well as those of Ukraine and the former Soviet bloc countries, it has quite a bit of work to do. If each person accesses the Internet just once a day for a month, like the Russians say, there will be at least 800,000,000 transactions to look at. Fortunately, there are tools available to make that job easier. There is too much information to not use them.

The Russians require an FSB monitoring device on each Internet Service Provider. The total monitoring done is in three programs, called the System for Operative Investigative Activities (SORM). SORM collects Internet traffic, mobile and landline telephone calls, and Wi-Fi and social networks, keeping the data for three years. Although it takes a court order to get data collected, the orders are secret and the courts have allowed data to be collected in cases where the target was a political opponent and had committed no crime. From this kind of database can come enough information to cause an individual a great deal of trouble.

A new legislative tool now makes getting the data easier.[26] Like the Chinese system of monitoring users, the Russians have started a registry of websites that is said to be aimed at child pornography. However, it turns out to be more than that. The Roskomnadzor (the Agency for the Supervision of Information Technology, Communications and Mass Media) collects information on banned sites from different government agencies. The sites that are blocked must be removed from access within 24 hours. If the Russians tell Twitter, Facebook, or Google that a website is banned, it has to be quickly removed.

Roskomnadzor has introduced deep packet inspection. This kind of tool is remarkable in the amount of information it can gather from a single transaction. It collects the payload (such as an e-mail), transport layer information (where it was going and the route it took to get there), the addresses involved, and the time at which the message components arrived.[27] It allows the user to see behind the message in the payload and collect various kinds of data from several types of communications. Over time, it can control user access, limit or allow user access to various applications, see who is leaking data to other parts of the world, audit a user, or collect forensic evidence for use in a trial. It can do all or part of this as needed, but (equally important) on a large scale.

The Chinese use a combination of filtering and storage solutions to manage information on a population of almost 650 million Internet users, 80% of whom are mobile. There are three elements to it: The Golden Shield, the Great Firewall, and the Great Cannon. These are concepts and may not exist in fact as a single place or thing, but every network in China is monitored and censored by a set of rules that are uniformly applied. The mechanisms are the same but the physical location may be different.

The Golden Shield is a data storage system that keeps records on every adult in China. It includes records from police, prison officials, state security services, border monitoring, population control, and an unending capability to find out where, and who, a person is. Greg Walton describes it as *a network that knows who and where you are.*[28] The main purpose of the Golden Shield is to keep a permanent record of who a person is by combining records from many different government agencies.

The Great Firewall is a censorship screening tool to look for potential trouble, screens some of it out, and blocks a list of sites that covers topics like Taiwan or Tibet. These are things that normal firewalls do. The Great Cannon is a new tool that can act as a government in the middle, looking for certain types of content and attacking the source of that content by injecting malicious code into it. That code can be used for tracking or monitoring individuals or the site itself. It can launch attacks against sites and effectively stop them from operating for some length of time. These processes are automated, fast and efficient.

The Chinese use a variety of technical methods to control information at the source. If there is a site that has information they want to restrict, they block it by restricting the Internet address, or else blocking the conversion of the server name (Domain Name Server controls) to keep users from finding it. In cases where multiple websites are hosted on one server, they block them all. These are not just sexually explicit websites but also those for news, health, education, entertainment, and certain religious groups, as well as a significant number covering many subjects on Taiwan and Tibet.[29]

In 2009 China went so far as to require the installation of software called Green Dam in all computers made in the country. It would have allowed monitoring and manipulation of data on any Chinese computer. The World Trade Organization finally ruled against the plan on trade grounds, but there are still 53 million PCs in China with the software *voluntarily* installed.[30] We have to wonder how many manufacturers are installing the software voluntarily in systems they export.

Combined with commercial tools for "law enforcement," any country can use data commonly collected on individual computers to geo-locate a person, find out what cell towers they use, create social maps of their contacts, track cell phone and Internet usage, detect their buying habits or when they visit (or attempt to visit) forbidden websites, read their e-mail and messages, map those into patterns of use, shut down their service, or limit their access. These tools can detect documents on a computer and match them to a list of restricted data. They can also check for similar documents, and don't have to rely on file names or hashtags being the same each time. If they are compatible with Google, they might use Analytics. Google tells me who accesses my website and when, what country they are in, and what browser they are using. Google knows a lot more than that, but doesn't always give what it knows for free.

Bruce Schneier, a popular writer on Internet security issues, elaborated on the types of data that various commercial businesses could use to get personal information about a person from the Internet.[31] We seldom think or know about all the information categories that can, and are, collected by our service providers. This data is taken in and stored by computers as part of their normal operation, and it is increasingly searchable by data-mining tools that examine quantities of data sold, or held, by various Internet and telecommunications companies like Facebook, Choicepoint, and Google. While reading this list, which I have added to, we might think about what governments can do with this type of information, especially when it is augmented by sources that only they control:

- **Location tracking data**: mapping tools and location records from GPS-equipped devices
- **Affinity, toll, and credit card data**: cards used by businesses to promote goods or services to customers, credit card tracing of purchases, or toll cards that trace the location of a vehicle or person
- **Wi-Fi and Bluetooth surveillance tools**: publicly available or enhanced scanning tools that show network connections, device connections like printers or scanners, and network traffic data
- **Banking transactions**: within-bank transactions by authorized users; between-bank transactions monitored by government financial institutions
- **Health information**: stored and shared between medical personnel at multiple places, from labs to insurance programs
- **Video surveillance and drones**: use of embedded cameras to show individuals' locations, activities, habits, or movements, and used with geo-location software (a computer's camera can be turned on remotely so that you, your home, or your business can be recorded without your knowledge)
- **Cell-phone surveillance**: conversations and location information are both recorded; complete voice communications can be collected
- **RFID surveillance**: tagged devices passing through monitoring equipment
- **Facial recognition software**: combined with other things like databases of photos, it can be used to find a context for an individual, like group membership or family relationships
- **Voice identification software**: recorded voices can be identified by matching a known profile with the recordings; the United States

says it used this method to compare intercepted voice transmissions of separatists in East Ukraine to identify the individuals involved
- **Public photos**: social media sites store millions of photos provided freely to others.

Of course, government data like driver's licenses; full body scans at airports; medical records and claims; passport scans; tax information; county and state records of land, incorporation, or vehicle purchases; real estate purchases; credit data; and the like could likewise be made available. Most governments don't have the resources to collect it all, let alone analyze it, but they don't have to. What they can do is get indicators that lead to a few people, and then collect a broader amount of data for those they have an interest in.

This data is stored on other people's computers, as in clouds, offered free to users. We, in effect, give our data to others to keep. Schneier sees this in terms of privacy, whereas users in Russia, China, and Iran might see it in the context of their personal security. SORM (and its Chinese equivalent) avoids all the difficulties of data collection by keeping the data in one place that is easy to access. Both systems control their Internet, though China is far and away better at it than anyone else.

There is no doubt that some countries see this data as something more than a commercial product to be sold. Authoritarian governments use it to control dissidents, political adversaries, activists, criminals, and journalists (which Green Dam could do by itself). An Amnesty International spokesperson said even setting up a social media group or a website can land an Iranian user in prison.[32]

Some of the applications or tools that are intended for monitoring crimes like child pornography, stolen credit card use, authorized surveillance on a warrant, or virus infections can be used for other purposes. These are powerful tools used to aggregate information that leads to a conclusion or decision. As an illustration of how that is done, we can look at the collection of data surrounding the murder of Odin Lloyd.[33]

William McCauley, Assistant District Attorney, testified in a probable cause challenge that the suspect, Aaron Hernandez, formerly of the New England Patriots football team, had been captured on video cameras; his phone used cell towers near the crime scene; toll records recorded his license plate when the car ran a toll plaza; receipts and video recordings from stores along the route showed the car's path; and telephone records indicated when phone calls were made to his friends, and who, in turn, called the victim. Listening to the stream of data that was available to law enforcement, it was apparent that there were sources of information leading back to the suspect. At least there were enough for probable cause to continue the case in court. Because they can be used for legitimate purposes, like collecting data for a murder trial, these types of software are not export restricted (as military equipment might be).

At the national level, governments have extensive capabilities that allow them to focus on particular individuals inside their borders. If they want to go across borders and tap into a computer, they can use another type of tool: the Remote Administration Tool (RAT). When it comes to attacks against individuals, this form is the most common, because it is also the most effective.

Dark Comet, IExplore, BlackShades RAT, and FinFisher monitoring tools, made by Gamma International GmbH, would be useful to those wishing to monitor computers. RATs are so common that they are the preferred attack method for most systems; they are available on the Internet and fairly easy to use. Citizen Lab, and several other security groups, give examples of how they work. The example below is a RAT called IExplore[34]:

A news organization operating a website reporting on China gets an e-mail that has an embedded story and four images. It also has two executable files disguised as images. A user who receives this file will see it as a story with attached pictures. Eventually, they will get to opening the two images that aren't, which run a program to install malware on the system, bring up a photo, and then erase themselves. The user sees a photo, just like he is expecting.

The malware has subparts that disguise the file so anti-virus programs will not detect it until after it has run. Another part monitors mouse movements and keystrokes, and allows monitoring of almost everything else the computer does. It communicates with Internet addresses in China and stores data related to the computer. In normal use, e-mail addresses and passwords are collected first, and then files that are of interest. The list of countries (including some like the United States) where this software has been found is lengthy. In the Citizen Lab analysis, there were 72 types of RATs that were identified. IExplorer was only one.

RATs get through the normal defenses of networks because they are not opened until they are already past the defenses. Businesses could stop allowing attachments to e-mail messages, filter e-mail at the boundary, or open each one in bounded partitions that would examine them for exploits, but those are expensive solutions that slow the speed of delivery. Some organizations don't bother with such security measures, but need to think more about it.

A more dangerous variety of RAT is represented by Dragonfly/Energetic Bear—an operation to get into energy sector systems, as reported in Symantec's blog in June 2014.[35] The RATs being used are *Backdoor.Oldrea*, customized software written by Dragonfly, and *Trojan.Karagany*, both trojan software that open doors to hackers to manipulate, use, and control a computer. They use phishing e-mails sent to targeted countries, similar to the way other RATs are installed and used. They are so successful that the amount of data they collect is too great to analyze. Symantec's report offers simplified explanations of how RATs work.

Sorting tools are needed because monitoring the old-fashioned way is too slow and cumbersome. Collectors of information, even those using common browsers and home computers, can get too much information themselves, and can be overwhelmed by it—a product of their success. Data becomes so voluminous that it can't be used without filtering.

The software that does that filtering, as a general category, is a data-mining tool. Data-mining software allows the discovery of previously hidden associations between variables that are potentially relevant for decision making.[36] Those relationships can then be categorized and reviewed for things worth analyzing further.

In the 1980s I managed a program called SHADOW in the Defense Department that looked at network intrusion detection. When this work was first started, much of it was manual and we were getting huge volumes of raw data to examine. The work was done by individuals trained to look at network traffic and make decisions about what type of relationships were indicative of a threat to the network. We didn't look at the content of messages—just the routing of them.

It was time-consuming and tedious work, but it was also reactive. It took weeks to figure out the kinds of attacks that were being used and how they were evolving. As soon as our detection identified a threat, the hackers moved on to some other technique that was not as well known to us.

We changed our approach to look for attacks that had not been seen before (today these are sometimes called Zero-Day Exploits). This new method produced far more interesting results, though fewer than would have been found by looking at daily intrusions. When someone at the U.S. Government Accountability Office says the Defense Department has 250,000 attacks per year, that number only reflects attacks that used known methods that can be detected and recorded. It's like the virus definitions that protect your computer; the virus software regularly downloads profiles of known viruses and looks for the familiar patterns. It is a lazy way to do business. New attack varieties thrive until they are discovered. When a new type of attack is identified, it takes time for software developers to modify the detection software to protect against the new threat and filter incoming data. The threat environment changes faster than the detection software; users are subject to being successfully attacked during those periods between a hacker deploying a new threat, detection of the new threat, and developing and deploying a countermeasure.

Ultimately, intrusion detection was automated so events, which are relationships between known activities on a network, could be categorized and shown on a monitor. Every day, we collected gigabits of data about those associations. A less trained analyst could then look at the trends to determine what actions had to be taken, depending on the importance of each one. That is what most of the monitoring tools do. Software is being developed that can identify new threats and produce countermeasures automatically. This works on new threats that use old patterns, but it is not useful for new threat patterns.

This means, though, that an extraordinary number of resources are required to check out all of the events that come to the analysts' attention, even though administrators don't try to monitor everyone on everything. Instead, they are looking for anomalies that allow them to focus on certain individuals. If we look for anomalous traffic inside a country, it takes resources to filter through all the possible events.

The Chinese used 1,000 censors for social media, 20,000–50,000 Internet police, and up to 300,000 low-level party members to check the content of Internet communications.[37] In defensive cyber operations we had two or three trained network analysts for small (by international network standards), interconnected networks, trying to find network intrusions. Where the Chinese have 650 million users, we had only a few thousand. We never looked at content, but we could make inferences from traffic analysis about what a person was doing. It took time and skill. It would take thousands of analysts for a population as large as Russia even if they had help from businesses that supplied services for Internet users. It is a very resource-intensive activity.

To make a censor's job more thorough, there are an uncomfortable number of cell-phone-monitoring software companies catering to businesses and individual families. They claim to offer the ability to monitor Facebook, Skype, Twitter, photos and videos taken by the phone, its location, messages, e-mails and such things as telephone number dialed, sender's number, message text, time of day, or what is on the screen of the user device at any point in time.

Most of these monitors are supposed to be for parents of minors who apparently need a great deal of supervision, but this goes beyond anything the NSA has been accused of doing. It is possible to block applications, uninstall them, or initiate a call from the user's device (presumably to call Mom or Dad). They work remotely, so Mom and Dad can watch while the kids are off at school. Teenagers may be skeptical of the need for this kind of software, but the numbers of companies offering it would indicate that their parents do not agree with them.

Along with parents, businesses do more monitoring than they officially acknowledge. There may be some twinges of regret at using a company phone where this type of monitoring software is used, but most companies don't advertise that they have it. Quite a few, along with many government agencies, use similar software for leak detection. When sensitive documents are leaked to the press or another government, the owner would like to know who did it. The newest software can help them find out.

One preventive measure that businesses employ is characterized by software that wipes a mobile device of sensitive data.[38] Fiberlink software allows a company to take sensitive data off the phone without the user's knowledge or consent, though legal issues have made cell phones a question. The devices can be company owned, or even one brought in to do company business and personally owned. We can imagine how nice it would be for a government to wipe the data from cell phones that were deemed to be offensive. Some won't have the same concern for legal issues.

We take for granted the notion that nobody other than parents or businesses is going to get hold of these types of software and use them for other purposes, but governments and criminals think little about using them in different ways. Researchers are working on ways to erase data after a predetermined period of time and some commercial applications have attempted to do that with e-mail, but the scalability and market for these types of tools are still untested.[39]

The Egyptians, before the Muslim Brotherhood was removed from office, used a tool called NarusInsight, named after the company Narus, owned by Boeing. NarusInsight was described as a tool that would allow deep packet inspection and content analysis of large quantities of data. Pakistan and Saudi Arabia had the same software.[40] Narus makes tools that take large data sets (big data) and analyze them for security teams to keep track of data in a company. The company's expertise is in sorting through millions of entries and picking those that need to be looked at. They provide tools to analyze and visualize this kind of data. NarusInsight and Bluecoat have similar abilities to monitor large groups of people.

Video surveillance is becoming more popular, especially among mothers wanting to watch their children. The most novel use of this surveillance is extortion through monitoring of home computers. Hackers have gotten into the cameras on home computers, recorded people in various stages of dress, and then extorted money from those who might want to hide what was recorded. We have to give them an "A" for innovation.

Video surveillance is much bigger than baby monitors and cameras on the front door of a home. This is a field that covers everything from a guard service using multiple video cameras at hundreds of sites (all motion activated and sound equipped) to huge drones orbiting large geographic areas sending back billions of bits of data for analysis. Homeland Security drones fly along U.S. borders; Hamas's drones fly over Israel; police body camera now record every move in an arrest. The problem for most forms of surveillance, aside from the legalities involved, is the quantity and redundancy of available data.

It isn't possible or necessary to look at all of the data, but "leaving it on the floor" can always be subject to second-guessing by a manager somewhere. Sometimes governments and companies keep it because they can be criticized for getting rid of what they no longer need. Simple filtering can be useful in this type of environment, but that doesn't solve all of the difficulties. Video footage is taken at different angles, under different weather conditions, at different times of day and night, and with different cameras, a problem also experienced by home security systems. Software developed by the intelligence services and defense contractors can compensate for most of these factors and stitch together images that look enough alike to find almost anything. This is useful only if we can afford it.

What governments can do that individuals can't is leverage multiple technologies across the entire spectrum of communications in order to exploit available data.[41] Satellite photos can be integrated with geo-located voice intercepts, shown with video of a truck following a route through a city. The truck stops, other photography, and telephone records can then overlay the network of people talking about what the truck is doing, making them into a social network. The social network can subsequently be linked to Internet social networks and mappings made to "friends" and groups this person belongs to. Computer records can show invoices and the company that rented the truck. Business records will show who the company and its officers are. If the event is something like an illegal drug delivery, the people who pick up the drugs can be identified and a larger network of intercepts grown until the dealers are identified. If this were easy, or cheap, there would be no drug dealers left in the world. So why are they still around?

Terrorists, criminals, and intelligence services have to develop counter-surveillance operations of their own to stay alive or out of jail. They use aliases that change often; switch delivery vehicles every few days; use prepaid phones, encrypted radios and networks; send their operations underground; and become overly cautious. One of my law enforcement teams observed a driver sitting in the parking lot of a hotel where the drug delivery person had just been arrested. The delivery man had a cell phone that he used to call the driver and ask him to come inside, cooperation with the police that might help him shave a few years off his sentence. The driver waited for a few minutes but didn't come in. We could tell he was talking on the phone to somebody, but we were hard pressed to say for certain that it was our delivery man. This driver had good instincts. He finally drove away and we were never able to identify him as the person who brought the delivery man to the scene.

There is one example of support for criminal operations that shows how protection from monitoring has become good business: the Russian Business Network (RBN). RBN is good enough that it is difficult to research in current open sources. The common anti-virus and security companies have good information; almost all have references that point to Wikipedia. Verisign, Symantec, McAfee, and Brian Krebs, the former *Washington Post* staff writer, all describe the organization in much the same way. We still don't know very much about RBN.

RBN is hard to find because of the nature of its work. It offers "bulletproof hosting" to anyone who can pay the price, which is about four or five times what it would cost if it weren't protected. (There is really no such thing as "bulletproof," which means a network can't be detected and can't be penetrated from the outside.) RBN probably is, or was, as close to secure as a system can get. In the language of special operations, this is a company of un-attributable communications—that is, nobody is really sure who owns or operates the organization, and its electronic paths can't be traced to the real owner. The members want to be bulletproof because they are hosting some criminals who work from websites and attack servers owned by others.

RBN, and similar service providers, claim to be legitimate businesses, which in Russia may be true. It supplies web hosting, software, and specialized tools to get access to information most people using legitimate hosting sites would have no need for. A report by the Shadowserver Foundation, a privately funded group of security researchers, provides details of what RBN sells.[42] It offers such things as password crackers, Trojan Horse software used to steal information, and a significant number of tools associated with criminal activity, especially the hacking of electronic bank systems. Shadowserver has traced RBN's operational footprints on the Internet and graphed the extent of the spread of some of its tools. In the past (and possibly still now), the operations end of RBN was "professional," meaning it supplied well-structured code and its applications worked well. It was, or still is, good at what it does.

Brian Krebs says as many as 60 percent of online cases of identity theft are the work of RBN, but the Russian government is not very good about helping law enforcement from other countries stop them.[43] Most public estimates say RBN's operations takes in $150–$200M a year, though those numbers are speculation, as is the organization's connection to the Russian mob.

Most sources put RBN's home in St. Petersburg, where Vladimir Putin is from, but its Internet addresses are not. In 2007, after some scrutiny and a series of articles written about where and how RBN supplied its services, it moved.[44] Even RBN needs circuits to operate and it has to get those from a provider. After the articles appeared, its circuit providers decided RBN might not be a good customer to have, and closed off its support. RBN tried setting up a front company in Italy and had its addresses moved there. The front company gave it cover for the addresses and the use of the circuits, but it was detected, and RBN packed up its networks and moved to China and Taiwan.[45] Since then, the organization is nowhere to be found.

There is speculation that RBN no longer exists. Some suggest it has been taken over by other businesses doing the same kind of hosting. We could just as easily assume RBN has a better front company that stands up to scrutiny. The mystery is good for business. The big cloud services, ISPs, and security companies have stopped referring to Russian hackers and now describe attacks as coming from "Eastern Europe" operations. It is more than just a curiosity. If RBN is still in business, it is good enough to not be traced to the country of origin, a kind of nirvana of hacking. The successful use of front companies seems to be their skill. If true, they have found the Holy Grail of hacking.

Governments use something even better that few individuals are capable of—specialized collection or attack software, the most well known being Stuxnet. Stuxnet is a worm, meaning it doesn't need to attach itself to something else in order to spread. In 2010 Symantec did a report on Stuxnet but recently found another version predates the 2010 version.[46] Stuxnet caused Iranian nuclear centrifuges to spin faster than their operational tolerances and ultimately fail. The attack was attributed to the United States and Israel.[47]

This kind of software is developed for a specific need, has to be sanctioned by the government that originates it, and has to be carefully constructed so as to leave no traces of where it came from. It may be used for collection of information, sabotage, or defense of a network.

Symantec discovered that the first version of Stuxnet was working three years before the 2010 version was identified, and it was somewhat different in what it was able to do. Instead of spinning the centrifuges, this variant closed a valve that was used to feed uranium hexafluoride gas into the enrichment centrifuge. It took a snapshot of normal operations and replayed the snapshot during the attack so operators didn't know the valve was malfunctioning. Masking the true result by capturing and replaying the snapshot was particularly clever.

Stuxnet and similar software is more sophisticated than most commercial software, and very specialized. It takes no imagination to look at the collection of disclosures by Edward Snowden and construct a profile of the software that must have been required to collect the different types of information that are described. Hundreds of programs would have been developed. Given the number of countries involved in spying on communications systems, and the types of software that are available already, it is hard to imagine a system anywhere in the world that doesn't have intelligence-gathering clutter in it.

Post-Crimea Russia is now clamping down on its own Internet while still allowing access to it. In 2011 the Russians were surprised by the influence of public media on the national election. Putin even said the Internet was a "CIA project." In 2014 Russia passed a bill to require any blogger with over 3,000 viewers to register with the state.[48] They will be subject to the same censorship as other mass media. (The CIA probably wonders where Putin got the idea that it had anything to do with funding or managing the Internet; after all, that was the Defense Department.)

In June 2014 the Russians met with Twitter representatives in an attempt to get a dozen "extremist" accounts terminated because they called for "illegal activity."[49] We can see a dilemma for Twitter here. Extremists are a concern almost everywhere, but the definition of what is extreme must vary quite a bit around the world. Which definition should Twitter accept?

Twitter is on its own, making rules that fit its business model. This is an area where Twitter customers may not even agree with their own government. Twitter doesn't get to decide who is right. Its policy says it will terminate accounts for certain reasons, and it tries to make the definitions broad enough to cover every situation. By being broad, Russia and China can find extremists who fit in those categories. Twitter has to decide, just as any other social media network does, whether to terminate those accounts or fight city hall. If Twitter fights, it runs the risk of being branded with the same iron as the so-called "extremists." Facebook and Twitter, being the largest companies, face this problem several times a day. Google is also up to its neck in Russian politics, although for a slightly different reason.

The Russian parliament passed a bill in July 2014 to require all technology companies to store personal data of its citizens in their own country.[50] Although the government can pass any kind of bill it wants, this one may have gotten through for reasons other than protecting the privacy of Russia's people. As a practical matter, it is not possible to do what the law requires. The real reason it was passed was probably to get Russia's Internet under control.

China and (to a lesser extent) Iran, Egypt, Syria, and a host of others have done a better job of controlling data inside their own countries, and they will get better still. This leads to concerns about the "balkanization of the Internet"—fragmenting it into pieces that are not public, filtering content, blocking websites of anything deemed "subversive," and using collected data to find disaffected citizens. China already does just that, and Russia sees the benefit of what the Chinese have done. Balkanization is already here, part of a trend the Internet society identified in 2007.

The concern is that Internet proponents see the service as something free and good. Like Google, it does no harm. An increasing number of governments don't see it as good, however, and want it controlled. Others want to exploit its access to other places as a path to information, which can be good or bad, depending on the reasons for gathering the information.

The Internet is like a major Wyoming road on which one can drive without seeing another person for miles. We used to drive over the speed limit, sometimes way over, because there was little enforcement of any kind. When there was an accident, it was usually a bad one. Cars would T-bone each other once in a while because the drivers were fixated on things miles down the road and bored to death. Stop signs were a nuisance, ignored, and point of impact for those T-bones. Cars would sometimes hit a moose or antelope, which did neither of those species any good, but did more harm to a car leaving the roadway. Fatalities were common. But we could get to where we were going very fast, so the risks were acceptable.

5. Cyberwar

Countries that don't like that kind of experience think the Internet allows individuals to take ideas and spread them around, giving access to other ideas in the process. Countries like Russia, Iran, Syria, North Korea and China want to control information dissemination, monitor the content, limit the access, and keep others from getting data from it. They want to control dissemination in the name of helping their people, by having them avoid things that might cause trouble. If you don't hum "Honky Tonk Women," life will be better for you. These countries want to define what is good and bad and what trouble there might be if the bad proves disruptive. They are helpful that way.

The Syrians are exploiting their access to find out what people are talking about; the Russians are less subtle. There are differences in approach, but the results are the same. People in these countries operate on behalf of their governments to monitor the content of their citizens' communications. But China and Syria have done better than that, going directly to the source.

In October 2012 the *New York Times* published an article that said Wen Jiabao, China's Prime Minister, had accumulated a fortune worth several billion dollars.[51] For the next several months the Chinese stole passwords and user accounts for most of the *Times* reporters and other employees. It wouldn't be too difficult to figure out what Chinese hackers might discover if they could log into the *Times* network, pretending to be the reporters who wrote the story on Wen Jiabao. It wouldn't take very long to find out who was providing information about his fortune. Those sources could then be warned that there is a cost to cooperating with the press somewhere else in the world. Admittedly, it is better than "We will kill you" threats made in Egypt. The Chinese are subtle, where others are less so. Pretty soon, sources will be hard to find.

China has successfully hacked a number of defense contractors, Bloomberg, the *Wall Street Journal*, and the *Washington Post*, so there are plenty of reporters concerned about their sources.[52]

Iran, one of China's best friends, seems to have adopted the same strategy. Whether the Iranians get help from the Chinese or just follow their success is another question. During the upheaval in 2009 China was said to have given support for monitoring telecommunications and crowd control to the Iranian government.[53] China is deep into other business areas of the telecommunications infrastructure.

Iran has had success in one of its first tries at cyberwar. In August 2012 it let fly with a destructive virus on one of its neighbors, Saudi Arabia. The attack was directed at Saudi Aramco, its largest oil company. It erased the files of three-quarters of the 55,000 employees, replacing them with a burning U.S. flag. The CIA's director said it was one of the most destructive cyber incidents ever.[54]

The year before, Iran began a longer-term operation known as Newscaster.[55]

The Iranians used personas on social media sites like Twitter, Facebook, LinkedIn, Google+, YouTube and Blogger to make contact with professionals in defense and foreign affairs. They became friends and corresponded with these people. They could find the usual social media information like groups they favored and places they visited. The operation used a fake news site to keep up their targets' interest. The news was plagiarized from other websites, something almost expected. The links were then used in spear-phishing attacks to steal credentials and insert other malware. In the end, the Iranians could steal almost anything on their friends' computers.

The Iranians are also said to be connected to Operation Ababil (previously discussed), an October 2012 denial of service attack against U.S. banks. They are learning how to use cyberwar techniques to damage an enemy, which in this case appears to be the United States.

If we see this as a function of war, it is another that can be fought without the participation of the militaries of the countries involved, though clearly whether militaries are involved is not known. Sometimes, hacking is a matter of how a country uses its military (and not what all military units typically do).

The Invisible Cyberwar

We can't look at conflict in the visible spectrum to see most of cyberwar. It is covert. People who couldn't talk about what they do or how they do it invented cyberwar. People who are good at keeping their work from public view perpetuate it.

Information war was mostly made up of secret things governments did to each other to win an advantage, but large parts of the doctrine were documented and public. Papers were written about it and policies were made to formalize its elements. Our military described it, in the 1980s, as a doctrine in seven parts: economic, command and control, electronic, intelligence-based, psychological, cyber, and hacker war. Taken together, these were total war.

Over the last ten years, we have begun to call much of this type of war cyberwar, even though it is probably not the correct term. We might have to coin a new term for it, since none of the existing ones fit very well.

Cyberwar was originally something else in the view of military planners. It was a small component of cyber efforts that focused on what computers could do to undermine the enemy. For a while it was confusing because the world spoke about cyberwar in one way and militaries spoke about it in another. Since that time, the focus has changed without greatly changing the components. The main difference lies in who does the fighting. It is a war fought less by militaries who claim its invention.

5. Cyberwar

In the 1980s we didn't have the best technology to work with to implement cyberwar. It languished on the shelves of most government agencies, particularly the Pentagon. It was called war because it was supposed to be used as an adjunct to fighting, to influence another country to do what we wanted. In those days, that's about all it could be. Militaries see cyberwar as an adjunct to a fight intended to bring another country around to our way of thinking—by force if necessary. That is an old-fashioned way of looking at it. Modern war has replaced traditional war, and military force, where it is needed, is now an adjunct to cyberwar. Overt fighting is a last resort, avoided as much as possible.

The military and the contractors supporting defense cringe at this idea, since they are not the ones who lead in much of what is a national security strategy. We give large parts of the battle to our intelligence community, so we don't need quite as many airplanes, tanks, and ships as we once did. Each country is somewhat different in how much they give to what part of the government, depending on how their agencies are structured.

As a simple illustration of the conflict, the Obama Administration has tried to get the CIA out of the business of drone strikes and replace it with the Defense Department. Drones fit into a part of information war that is not cyberwar, a part called intelligence-based warfare. The intelligence collection, analysis, and targeting are all done by computer, but so is targeting on a tank or flying a combat aircraft. We don't call the use of tanks or aircraft cyberwar, though maybe someday we might.

Not only are militaries fighting fewer wars, but wars are also being fought covertly, using military weapons and capabilities but not soldiers in the traditional sense. In the United States they are fought by intelligence agencies like the CIA, the NSA, and the FBI, and other agencies not part of intelligence, like Homeland Security, the Internal Revenue Service, and Treasury.[56] The military supports the roles that others increasingly play in fighting. Yet some of those areas of war were *never* in the purview of military forces, with economic warfare being the most prominent.

6

Economic War

According to military doctrine, economic warfare is the manipulation of information exchanged in trade (either denial or exploitation) as an instrument of state policy. We might ask how that aspect got into military doctrine when it is outside the scope of traditional military activity. The U.S. doctrine on economic war implies that the United States might be willing to use its capabilities toward this end and, in rare situations, has done so. Economic war is the primary component of information war, but it is far from the way the military described it when it was doing the fighting. Cyberwar has redefined economic war into a major element of modern warfare.

The three main uses of cyber in economic warfare are theft, denial, or manipulation of data that is used in commerce. The world press seems to focus on the theft and forgets that there could be a good deal more to it than that. The greater harm comes in manipulation. In 2015, the theft of 21 million personnel security records from the Office of Personnel Management was mostly reported as a theft of data when the greater risk was for an intelligence service to add records, modify, or delete them. It is possible to create records of people who will now appear to have a security clearance, modify them to show "adverse information" like drug abuse or child molestation, or remove them so a Federal agency is not able to verify the background of an applicant. This delays or denies consideration of an applicant or gives someone who was never an applicant the appropriate investigation for a clearance.

Dragonfly/Energetic Bear are names given to the activities of a group that is doing more than simply stealing data.[1] Dragonfly was a group of hackers who started hacking U.S. and Canadian defense contractors sometime around 2011 and moved away from that to hacking energy grid operators, major electrical generation firms, petroleum pipeline operators, and industrial control systems in the United States, Spain, France, Italy, Germany, Turkey, and Poland.

This group specifically targeted makers of software used in industrial control systems, modifying that software while it was posted on websites for distribution to field units. The infected companies became the delivery mechanism to their customers. Symantec was not able to identify the originator beyond "a country in an eastern European time zone," but it represents what is possible when a country concentrates on a single market sector and attempts to find out all it can by penetrating its human users, and then undermining its very technology fabric.

Whoever ran the Energetic Bear operation can see any document a user prepares or on which they comment. They can anticipate the types of responses to the attack that will occur if it is discovered. They can control substantial amounts of an industry that controls energy sources for more than one type of energy. This group is thinking big, and Symantec says it looks like the members have government financing behind them. We don't have to look very far to see what governments those might be.

This is the potential of economic warfare to go after the manufacturing and distribution systems of products and services. This is a level above the mere theft of data, which may also be involved. These kinds of attacks are able to undermine the fabric of a country or a worldwide industry sector. They are thinking big.

In 1997, a Swedish teen used his computer to hack into a Worcester, Massachusetts, telephone system and knocked out 600 users in the community. At the same time, it brought down the 911 emergency system and kept incoming aircraft from activating runway lights at the local airport. The Secret Service investigating the case found 22,000 telephone switches that could be disabled with 4 keystrokes.[2]

In May 2014 the U.S. Justice Department indicted several individuals who work for China's People's Liberation Army, announcing in a public release these government employees were hacking industries in the United States. They stole proprietary information and trade secrets that helped the Chinese leap ahead in product design and release. In China, business and government (including the military) are close to being a single entity, unlike most places in the West.

In a way, indicting a hacker in China for activities they carried out while they were living in the United States is pointless. It might give the Justice Department a warm glow, but it won't accomplish very much. These spies are not coming to the trial. The indictment is interesting because it means that, if there were to be a trial, the prosecution could prove some of the things the defendants are accused of. If for no other reason, that part of the process is important. It is attribution that is important in laying blame.

As in any case like this, the hacker defense is always "It wasn't me." They say somebody may have hacked the system, but our security was not very good and the hackers who did it probably found it easy. Somebody may have used the defendant's system to get in, and he feels bad about that, but it wasn't him; his security needs to be better too.

Attribution is the ability to say, within the standards of reasonable doubt, that an attack came from a particular place, entered the target system, and bypassed safeguards designed to protect the system. Almost everyone in security these days will make general statements about the source of an attack, but they will not be too specific. As Symantec did with Energetic Bear, they say "Eastern European sites," when at least some of the code suggests a Russian source. Most businesses say the same thing when describing where an attack originates.

Hackers are smart and use "hops" to hide. They get their hops by hacking some little furniture company in Iowa and using it to hack into other businesses, but they could hack a small business in Spain, use that to hack a research facility in Switzerland, and then use that to hack the furniture store. Hackers also use "bounce sites," where a server's only purpose is to mask the next hop, provide a fake path that leads nowhere, or extract embedded content from what appears to be normal traffic and forward it to another hop. Finding the source gets harder with each hop or bounce site. Shadowserver found some using seven or eight.

Stealing information that will be used to manufacture new products is a long process, although the kinds of things being stolen can affect the timeline. If the attacker is stealing designs, many new programs will link designs to fabrication by computer-controlled machines. When they steal designs, they are stealing the programs to automate the manufacturing or fabrication processes carried out by Computer Aided Manufacturing (CAD) systems. They can speed up design and manufacturing in one stroke. Some countries don't have the patience or the wisdom to do their own work. The Chinese feel like they deserve whatever they can get; so do the Russians. When either one of them is stealing designs of the internal networks of a city or a major business, they are probably looking for a way to disable, or get into, several at once. That prospect should disturb us.

Attacks against economic sectors are more common today than they have been in the past. The Iranians tried to attack the banking infrastructure of the United States after sanctions were tightened against them in 2013.[3] So we know they want to cause trouble. Their goal was to slow or stop the ability of banks to do online banking. The Iranians started by amassing computers around the world to launch a denial-of-service attack. They managed to hack data centers and cloud services that were then used to launch the attacks. How they got control of these data centers and cloud services (and which ones) is the missing link in the story.

The attack was bigger than what Russia used against Estonia in a 2007 incident that lasted over a month. During that time, Iran went after some big banks: Bank of America, Citigroup, Wells Fargo, U.S. Bancorp, PNC, Capital One, BB&T, and HSBC. The CEO of PNC, on CNBC's *Squawkbox*, said the attacks originally came from Russia but changed location several times after being detected. They did slow down the ability of users to access their accounts, but didn't harm the U.S. banking structure. That isn't to say they couldn't have, though.

What the Dragonfly/Energetic Bear operation demonstrated was the willingness of a government to undermine the software integrity of a whole industry. That could have been banking, but it wasn't. Among other things, we should probably be looking at the software driving the analysis engines in intrusion detection and virus software to make sure it hasn't been tampered with by something like Energetic Bear.

The concern has evolved into something called supply chain protection. Vendors of software and hardware have to know where their suppliers are coming from and the distribution chain that gets the product to a customer. That means protecting the source software at every stage of development, sometimes even looking at equipment on a customer site and comparing it to what was manufactured.

That is also part of cyberwar. The success in the electrical grid is probably already spreading to a number of countries that will start using the same techniques in different parts of the infrastructure. In several years, there will be a significant increase in infrastructure attacks that will be complex and difficult to counter. It would not be too hard to imagine a Stuxnet in the banking system opening or closing accounts of real bank customers or telling ATM users that the machines are out of money. Using the same attack methods as credit card thieves, hackers could destroy credit records instead of stealing them. We have that prospect to look forward to unless we figure out how to stop them.

Governments don't have to make money like credit card thieves; they can cause interruptions in service or damage that cost time and money—time and money not spent on making a profit. They do have to worry about retaliation in kind, however. When they look at the electrical grid, gas and oil infrastructure, transportation systems, banks, financial institutions, emergency services, and telecommunications as legitimate targets, they have to consider the consequences. One of the reasons governments hesitate to act is their uncertainty about other countries' capabilities to do the same to them.

7

Economic Espionage

Economic espionage is part of the broader area of economic warfare. When countries steal from each other with the idea that their economies will gain from it, they are engaged in espionage. The relative strengthening of one economy over another is both a political and an economic incentive.

U.S. intelligence services are prohibited from stealing information from another country and turning it over to businesses to give them a competitive advantage. It is part of the policy of our intelligence community that intelligence is used to establish facts and the accurate interpretation of facts, but not to engage in manipulation of the economy.

It is on the wall at the entrance to CIA headquarters: "And ye shall know the truth and the truth shall make you free." It is idealistic to some, but realistic to those who do the work. The CIA can't steal from Chery and give that information to GM to get an advantage on a bid for new fleet cars. Knowing the truth—that the Chinese would not act the same—does not help us very much.

The Chinese once had scientists flying to the United States, going out into the corn fields of Iowa, and stealing all the seed corn. DBN, a Chinese conglomerate, has a seed-corn business and it was trying to improve the quality of its products. A total of seven people were indicted; two people were arrested and charged with stealing genetically altered rice.[1] The Chinese will steal anything and deny doing so even when they are caught red-handed.

We have an antiquated view that our intelligence services are to be used to advise government leaders on what the rest of world is up to. The U.S. belief that facts will change the way policymakers work is an honorable idea, but one rarely borne out by history.

Almost every major business and every country has an intelligence function, and all of them use computers for a large part of these activities. When Willie Sutton said he robbed banks "because that's where the money is," he had never seen an ATM machine. He took a lot of chances to get his cash, but now he could do it from home, with a cold beer in one hand. Cybercrime is much safer and a lot more rewarding, but not all of this activity is criminal.

Stealing business information has been around for hundreds of years, but computers have made it easier to do. A recent national intelligence estimate said four countries are the most active in cyber-espionage against U.S. targets, and two of them are friends: China, Russia, Israel, and France.[2] Economic war is driving the hacking of businesses, but stealing proprietary information to get successful fast is far from new.

France has always been known for its industrial espionage, and openly admits that it spies. We admire the honesty of the French, but at the same time, this knowledge made their criticism of the National Security Agency seem hollow. The French seem a bit schizophrenic about it.

Robert Gates, in describing the French use of industrial espionage, was as blunt as any former Director of Central Intelligence has ever been:

> I said for years, the French Intelligence Services have been breaking into the hotel rooms of American businessmen and surreptitiously downloading their laptops, if they felt those laptops had technological information, or competitive information that would be useful to French ... companies.
>
> France has been a mercantilist country. The government and business have operated hand-in-hand since the time of Louis XIV. This is not exactly a new development in France.[3]

Some of France's most famous escapades, like putting microphones in commercial aircraft to spy on IBM and Texas Instruments, are renowned among intelligence services.[4] At the Paris Air Show in 1993, France targeted a new Pitch Yaw nozzle on a Pratt and Whitney jet engine, causing several U.S. firms to cut back on their displays and limit exposure of some of their aircraft. Yet the French are astounded by accusations that they spy on anyone. Well, possibly not astounded, but certainly surprised to hear it said out loud by another government. Objecting to what the NSA has been doing is almost sanctimonious on their part.

There was always an unwritten law that intelligence services didn't make these kinds of things public. By Jack Devine's (the former CIA Chief of Operations) account, the French were running black bag operations (i.e., breaking into the offices of business executives in Paris, copying their internal documents, and leaving little trace). In retaliation the CIA Director said, "No more Mr. Nice Guy," and started recruiting French government officials who had access to trade secrets in entertainment and telecommunications. The French found out what was going on and had four of the CIA's officials removed from France and made the whole effort public.[5] Eventually, all of this kind of foolishness goes back into the darkness where it belongs.

In 2011 the Office of the National Counterintelligence Executive published a report to Congress outlining a few of the public cases noted to be from China:

- "In a February 2011 study, McAfee attributed an intrusion set they labeled 'Night Dragon' to an IP address located in China and indicated the intruders had exfiltrated [stolen] data from the computer systems of global oil, energy, and petrochemical companies. Starting in November 2009, employees of targeted companies were subjected to social engineering, spear-phishing e-mails, and network exploitation. The goal of the intrusions was to obtain information on sensitive competitive proprietary operations and on financing of oil and gas field bids and operations.
- In January 2010, VeriSign iDefense identified the Chinese Government as the sponsor of intrusions into Google's networks. Google subsequently made accusations that its source code had been taken—a charge that Beijing continues to deny.
- Mandiant reported in 2010 that information was pilfered from the corporate networks of a U.S. Fortune 500 manufacturing company during business negotiations in which that company was looking to acquire a Chinese firm. Mandiant's report indicated that the U.S. manufacturing company lost sensitive data on a weekly basis and that this may have helped the Chinese firm attain a better negotiating and pricing position.
- Participants at an ONCIX conference in November 2010 from a range of U.S. private sector industries reported that client lists, merger and acquisition data, company information on pricing, and financial data were being extracted from company networks—especially those doing business with China" (Mandiant, 23–24).

The Chinese are far and away the best and most persistent of thieves when it comes to economic espionage. There is one report that sheds a little light on this subject: "APT1: Exposing One of China's Espionage Units," by Mandiant Corporation, a FireEye, Inc., company that does Internet threat detection, tracing the threats back to their electronic hosts. Mandiant helped the *New York Times* find out that it had been hacked by China, particularly a group of soldiers operating from an address it had seen before. Several times, smaller security companies were the ones making headlines in following and tracking hackers around the world.

On those rare occasions when the Chinese can be monitored for a long period of time, which Mandiant seems to have done, there are a lot of things that can be learned about their operations. Mandiant claims to have been following the PLA soldier group since 2006. In Internet terms, that is a long, long time.

7. Economic Espionage

Watching hackers over time has great benefits. The problem is whether the operation can afford to permit the losses of information while the hackers are being followed. From 1998 to 2000, the group I worked with followed one small group of six people, and they were able to get into almost every network they tried because they were persistent and had quite a bit of time on their hands. We were interested in how they did it, but conflicted over how much of their activities had to be reported to those being attacked. We saw the hackers use brute-force attacks on password files, store their information on another computer that was hacked for that purpose, change the locations of their attacks so they appeared to come from other countries, and collect tools they used to build software. They would test their software and improve it before applying it to systems they were attacking.

Some attacks were short and never repeated; some lasted for over a year. The hackers were not very advanced in their techniques at first, but they got better as time went on. Eventually they came up with a way to send a large number of requests to a website so that it would become overwhelmed or slow down the service; they were launching these attacks from other computers, not their own. This was denial of service on a larger scale than we had seen before. Our bosses felt that this was too big to allow and went to law enforcement to shut the group down. Keeping hacker-tracking secret, once it is elevated to senior management, is difficult to do. Eventually, somebody has to be arrested, disrupted, or stopped because the potential damage is too great.

Police get into the same situation when running a sting operation. Suspects bring in car stereos, laptop computers, jewelry, cameras, and almost anything else that has cash value. How long do we let that go on? Are we part of the problem if we pay them for these goods? Do they steal more because we are there? Eventually, we see enough of the same faces, and someone higher up says, "Shut it down and arrest them."

In six years, as Mandiant found, hackers can do more damage than anyone could imagine, especially if they are organized and have money behind them. In one instance Mandiant observed the PLA unit stealing 6.5 terabytes from a single organization over a 10-month period. That isn't very much information, but the quality might have been good.

In another case, Mandiant observed a penetration that lasted two and a half years and was reading the e-mail of the President and General Counsel of a company. During that time, the Chinese negotiated a much lower ("double-digit") unit price on goods purchased from that company. This indicates they are not only stealing but also plowing the stolen information back into the economic war they are fighting with us. They were stealing information from multiple companies, worldwide but mostly in the United States, and from different sectors of the economy:

Information Technology	Transportation	High-Tech Electronics
Financial Services	Navigation	Legal Services
Engineering Services	Media/Advertising	Food and Agriculture
Satellites and Telecoms	Chemicals	Energy
International Organizations	Scientific Research	Public Administration
Construction and Manufacturing	Aerospace	Education
Healthcare	Metals and Mining[6]	

In all, Mandiant observed hacking at 141 different targets in 15 countries. On average, after penetrating a system, the hackers stayed inside the network for over a year. They had a large infrastructure to support this kind of theft, with 109 servers in the United States and various single-digit numbers in 12 other countries (13, if you want to count China). They worked out of a building that was 130,000 square feet and had hundreds of employees. We remember the Justice Department indicted just a few. They didn't need 109 servers if the only information they were gathering came from the 141 systems Mandiant was monitoring; the number of employees seems inordinate for the number of systems they were hacking.

Keep in mind, Mandiant is only reporting on one active hacking organization. It recognizes there may be some that have been missed.

The ONCIX report, without citing examples, notes Russia does just as well. That is a little surprising, since the Russians are as good as the Chinese and there are many examples of their handiwork, starting early on with Moonlight Maze.

The U.S. government has a penchant for giving names to events so it can keep track of a series of intrusions without having to reference each one. Moonlight Maze was a three-year series of Russian electronic collection attempts, for the most part directed against the U.S. military. Similar collection attempts by the Chinese in 2004 were called Titan Rain.[7] These names help organize and summarize incidents lasting several years and come in handy when one like Solar Sunrise turns out to be three teenagers and not a foreign government.

Moonlight Maze caused a stir because it was identified as coming from Russia and directed at parts of NIPRNET, the Defense Department's unclassified business network.[8] From all indications it was military on military, and it probably qualifies as cyberwar, in the current sense, and not economic warfare. However, it does show how the Russians protect their intrusions into another country's networks.

Moonlight Maze occurred early in the existence of the responders, like the National Infrastructure Protection Center (which was still literally unpacking boxes when I went there) and the Joint Task Force-Computer Network

Defense. They were too new, and not well enough equipped, to manage these types of attacks. The military units proved inadequate at stopping them. There was a tingling sensation on the backs of necks in the higher levels of the Pentagon.

In 2000 the United States sent representatives to Russia to offer assistance in identifying the origins of some of the attacks. The U.S. representatives provided telephone numbers; after waiting several weeks to respond, the Russians finally said the phone numbers did not, and never had, existed. "No hacking coming from here," they claimed. U.S. government officials later blamed the Russian Academy of Sciences for the attacks, but the Russians denied it to the end. We should expect them to, since everyone who initiates an attack denies doing it.

The Russians and Chinese use their intelligence services and militaries to collect information from the competition and feed that back into their businesses. So do the Israelis and French. The Iranians likewise do the same thing, though more out of a necessity to work around sanctions. The Russians were doing this a long time before the Chinese got in the game, but the Chinese don't do anything small after they get going. The Chinese are stealing intensive amounts of our proprietary business information and relaying it back into their industries as a part of a national strategy.

A friend of mine lived in China, near Beijing, for five years while her husband worked for a large mining company. After reading my first book about Chinese information war, she observed that I had overlooked a cultural aspect of the Chinese stealing from the rest of the world: they believe that if someone does not protect something as if it is valuable, they deserve to lose it. If their businesses need information and that information can be stolen easily from us, we have no right to it. Because we are careless, we deserve to lose it. We need to play the same game because this attitude, which we find offensive, is not offensive in their culture; convincing them that stealing our proprietary information is not acceptable behavior requires a will to make it unprofitable. U.S. administrations have thus far not demonstrated the will to make China pay for its transgressions, allowing it to become a cost of doing business.

In a way, the Russians have the same strategy. For most of my government career we thought the hackers of the world lived in Russia. When an attack came, or a new one was discovered, it was always from Russia. That isn't as true today, but the Russians do operate some of the largest hacker operations in the world, and they have operated longer than any we saw anywhere else. Most of them are criminal, intermingled with legitimate businesses, and tied to the government.

My brother worked in Moscow trying to help the newly formed government refine its legislation process. The Russians apparently thought they needed to be able to generate, track, and manage their system the way we do in the United States. He first encountered large black cars parked along the curb outside government offices, casinos, hotels, and other locations around Moscow. Some days they were double-parked. Anyone in Washington, D.C., sees the same kind of thing now and again, but these cars have diplomatic plates. Our police will move them along pretty fast. The Moscow occupants, however, just got out and left their cars in the street, generally with "attendants."

My brother thought these individuals were government officials who were getting special care, but one day, walking with a Russian co-worker, he asked about them. "Stay away from the black Jeep Cherokees," he was told. "They are the Russian mob, government officials, or special people who are friends of the government. It is hard to tell them apart. Just leave them alone." The statement was very matter of fact. A friendly reminder, but worth knowing. It was difficult to distinguish between political leaders, mob bosses, and the oligarchs, even for people who live there. The Russian business leaders are intertwined with the government and organized crime in ways that are unique to them. (My brother added that he never had an unfriendly conversation with any of the attendants or occupants. They were all outgoing, helpful, and friendly—even the Russian FSB.)

The United States and most of its allies can't organize themselves in the same way the Russians do because we pretend business and government should be separated, even though they aren't. The main difference is that free world companies usually aren't directed by the government, or managed by them, though there are variations of that theme even in the United States. But these are not government-managed economies where quotas and business deals are made in government offices. Businesses have influence, which they usually pay for on both sides of a transaction, though the relationships are tenuous, fluid arrangements between leaders in business and government. We know the political affiliations of Google without doing a lot of searches to find out.

The Chinese have government political offices in their larger businesses and, as the *60 Minutes* crews found at Huawei, government officials do their best to hide them from outsiders.[9] To complicate matters, the army, the intelligence services, or some other part of the government sometimes owns their international companies. They deny any such thing, and swear by the official policy that bans military leaders from this kind of arrangement. But the officers whom the Chinese government charged with living beyond their means are getting that money from somewhere.

Russia has a different structure of business, with oligarchs owning almost everything but remaining dependent on the good graces of government officials to keep their status. We saw what happened to Mikhail B.

Khodorkovsky, oil billionaire, when he decided to challenge Putin. He was lucky to spend only 10 years in jail. He was a vivid lesson to his contemporaries, and notice to the rest of the world, that Russian businesses are not quite the same as those in the free world. Either you are a friend of Putin or you are not. This is not like being left off the guest list at the Inaugural Ball in Washington; this is going to jail or getting beaten up. There is a very distinct difference.

Western militaries cannot fight this kind of economic war. Most of the time they are on the fringes at best. The navy rescued a ship's captain taken by pirates from the *Maersk Alabama*, the army fought a losing battle of interdiction of poppies in Afghanistan, and the air force flies cover for commercial convoys going to the interior.[10] In most countries it isn't in military charters to do much more. They have a role to play in maintaining the freedom of the seas, skies, and outer space where satellites live, but not in business relationships, outside of government defense spending (where their influence is considerable).

Russia and China would like the world to believe that every country spies and the United States spies more, and better, than most. This is not exactly the same thing as economic espionage. They use Edward Snowden to portray U.S. intelligence services as collectors, with the rest of the world being wholly ignorant and innocent of the same types of activities. They structure these arguments to suggest that the United States does the same kinds of things to extract information from business and government, and does it better than any other country. This odd logic suggests the Russians and Chinese will formally deny any such activity, but admit to not being as good as the United States. They would be better off not commenting on it. They have changed the rules of the spying game by making classified information public, and they will regret it one day.

Russian news stories imply that the United States will give intelligence to U.S. businesses.[11] They are using a trick of logic that works both for and against those evaluating the cause-and-effect relationships between things they see—situational logic.[12] Most readers don't have the time or inclination to analyze things, so they apply their own bias. Facts in evidence do not always make the case. Edward Snowden was a spy, like many others before him, who worked for Russia, but the party line is that his idealism made him do it. In fact, he was a highly privileged administrator of computer systems who stole quite a bit of very sensitive information and made it public through the U.S. and UK press.

We can't stay on even ground with other countries, some of them allies of ours, when their intelligence services are allowed to steal secrets and use them for economic advantage, while ours can't. The United States can't fight this kind of war on such uneven ground.

8

THE EVOLUTION OF COMMAND AND CONTROL WARFARE

Command and Control Warfare

Information war doctrine grouped under "command and control warfare" (C2) the techniques used to attack the enemy's ability to issue commands and exchange them with field units. What this really means is finding out who runs something and using information coming from that person to influence others. There is nothing unique to the military in doing so, except that military units run their operations against other military units. In a global war, that concept is obsolete. C2 takes on a major new role in cyberwar.

Nothing demonstrates the evolution of C2 Warfare more than drone strikes, which are largely not done by the military. The principle is the same as the military envisioned, but the military is not flying those drones. Second, C2 Warfare is being used to select leadership in other countries and influence those leaders to change positions or follow the lead of another government. This manipulation is accomplished by cutting off the press sources for stories in other countries, tamping down internal dissent related to the political objectives, sorting out who is on what political side, limiting or controlling television and radio stations, disrupting and denying telephone service—all these are tricks that come from knowing what your enemy will do, when they will act, and under what circumstances. The Russians didn't focus on everyone in the Ukraine. They didn't have to.

The Russians were in the Ukraine with a monitoring tool, called "Snake" by some, but Agent.btz by people in the computer security business.[1] This tool is difficult to detect with traditional anti-virus software. Snake communicates with the outside world to give access to internal files accessible on the infected computer. Most of the samples analyzed come from the Ukraine, where about 32 instances were found, the most of any country. The second highest number was in Lithuania.[2] Those occurrences were mostly in government offices. It wasn't detected until 2006. Since Snake's discovery, there have been many "improvements," making it harder to find. The Russians say, "It isn't us."

This tool is a simple and effective way to gain access to a victim's computer and the environment surrounding the computer. The user (that is, the person accessing Snake on the victim's computer) can read files, e-mails, and anything produced on the computer. He can use it to see into the office or home of the victim by activating the computer's web camera. He can also listen by activating the microphone. This type of collection can help the user decide who will favor his government and who will not. He can look at plans and actions the other government is going to take, and factor this into future planning. His people will know who to count on, who to influence, and who cannot be influenced.

Command and control warfare has become more than just military on military. It is government against government, where each is trying to influence the other into sharing a common belief about a policy or behavior. The Internet and television are the methods of delivery; the weapons of persuasion are real. They may not be as effective as many governments would wish, but they are more effective than traditional ways of influencing with mass communications. Both Russia and China still take out large ads in local newspapers and magazines extolling the virtues of their homelands to potential tourists and business leaders who might want to open plants there, but they are far more effective at using controls on their own press.

Business leaders are as much a part of war as governments. We now have business leaders meeting with heads of state more often than their own countries' leaders. Their business areas are vertical, and directly related to how governments fare in global markets. They become part of the ability of one country to influence the policies of another.

Electronic Warfare

Electronic warfare is intended to enhance, degrade, or intercept radio, radar, or cryptography of the enemy. This is an old, very sophisticated, and highly classified field that has had more success than most other areas of information war; yet almost nothing is known about how it is done, or where. Cyberwar has not really changed what we know about electronic warfare. In the Second World War we got a glimpse of how this works with the breaking of Japanese naval codes.

There were two sets of codes—the Flag Officers Code and JN-25, a more widely used code. The Flag Officers Code was never broken, but JN-25 was. It had over 27,000 entries augmented by a second codebook, which had 300 pages of numbers, sequenced by 5 digits. The navy collected messages, even though they knew they couldn't decrypt the code. They saved the messages while they looked for the keys to the second book, which they eventually found. By the time they figured out part of the second codebook, however, the codes had been changed and Japan had attacked Pearl Harbor. With British help, the United States was able to uncover parts of the new code, and combine it with what the British already knew.[3] Eventually, having the codes paid dividends. But it is not hard to see why the work was classified for so many years. When we have someone's codes, it gives us an advantage that we don't want to expose.

One of the strangest stories to come out of World War II was that of Stanley Johnson, whom most people have never heard of. Stanley was an embedded reporter (before they called them that) from the *Chicago Tribune*.[4] He apparently stumbled upon an intelligence file, presumably a classified one, while traveling with the Pacific Fleet. It gave him the idea that our navy had broken the Japanese naval codes and knew in advance what they were going to do. This was in 1942, at a time when the war could have gone either way in the Pacific and Allied lives were at risk in a number of places.

Johnson wrote the story, apparently got it past censors (who, in those days, were supposed to be on watch for this type of thing), and the *Chicago Tribune* published it. The *Washington Times-Herald* later published the story as well. This is freedom of the press, something that has come back to haunt us repeatedly in recent espionage cases.

According to the analysis done by the Justice Department, the law was on Johnson's side, though the same type of teeth-gnashing done by the Justice Department today made James Rosen, from Fox News, a co-conspirator in a case of apparent espionage (that case was nowhere near as dangerous as our knowledge of Japan's codes). Is giving the story to the *Chicago Tribune* to run, knowing the consequences to our naval forces in the Pacific, something other than journalism? I guess we had to be there to understand the greed that allowed such a story to be published. The *Chicago Tribune* editors must have thought about it a little before they published the information. The government brought the case to a grand jury but was unable to obtain an indictment.

We need to reexamine the responsibility that journalism has to national security, and put law behind the limitations that such a reexamination would imply. If we are to have a free press, it has to be responsible in its reporting. Journalists, publishers, and bloggers frequently use the First Amendment to make public some of our most sensitive secrets. Publishing such secrets puts the country, our allies, and our agents in great danger; the information is often deadly to agents and friends. Almost no other country in the world allows it.

Intelligence-Based Warfare

Intelligence-based warfare is the integration of sensors, emitters, and processors into a system that integrates reconnaissance, surveillance, target acquisition, and battlefield damage assessment. In cyberwar, this area is not much changed from its original inception in information war. We see drones every day that have more intelligence power in their noses than some of our older satellites had. They give us better pictures, faster, and in volumes that stagger the imagination. They are supposed to be integrated into a system that allows those images to be transmitted quickly, analyzed, and acted upon. The whole system assumes we can get ordnance on a target almost immediately after the analysis identifies a target.

We could see how well, or poorly, that works in April 2014, when Al Qaeda had a meeting in Yemen that was outdoors and big enough to attract attention. Somebody actually produced a YouTube video of the event. It introduced Nasir al–Wuhayshi, who spoke to the group and is one of the suspected successors to Ayman al–Zawahiri, the long-time number two, and now number one, leader.[5] A few of our press corps military experts, and quite a few bloggers, wondered why someone didn't drop a big bomb on this gathering, as it could have solved a number of problems all at once.

YouTube uploads 100 hours of video every minute of every day. People watch 6 billion hours of it every month.[6] That range makes a single video hard to find unless someone knows where to look and, in this case, when. This video wasn't released until the meeting was long over, and some of the faces were electronically obscured. Even though it was on every news broadcast for the day, we can't be sure there ever was such a meeting, or that it occurred as described by the video. Its value as a credible information source was not very high.

Most of the world is watching for meetings of people who are Al Qaeda, but only a small number of those watching can hit the group with airstrikes. Quick airstrikes usually mean an asset is already at the site, flying around waiting for something to come up that requires action. Intelligence-based warfare is supposed to help us place assets where they are needed, but the assets can't be everywhere. Anyone who saw *Zero Dark Thirty*[7] knows this is not as easy at it sounds. People hide from aerial surveillance. A lot of moving parts make it dangerous and complicated.

Once we find a potential target, we are supposed to have the capability to "seamlessly" collect data about the target, analyze the data, call the drone or aircraft strike, and check for damage after the fact. Defense contractors make this sound easy if the government buys something they make to simplify the process. They tend to pretend the data will be collected, processed, and analyzed, and a target package will pop out in a matter of seconds. The hard part is that all of these things have to happen perfectly, and, in addition, we want to make sure there are no nearby wedding parties or public schools in the target area.

In April 2014 a drone strike killed an Australian and a New Zealander in Yemen.[8] This complicates the process even more because these are allied countries. Al Jazeera says somebody blew up a wedding party in Yemen by mistake, but nobody is saying who it was.[9] We know the Taliban will call a gathering a wedding party even if it isn't, so the confusion is great.

The meeting can be anywhere. Al Qaeda says it took place in Yemen, but that may just be a story that will make us look all over Yemen for the site, and scour through lots of pictures of every open space there. It can be somewhere we can't see or control very well, or someplace where our bombs would not be welcome. We could not find a commercial jet missing for months when it was trying to hide, and those are big. It takes time to pinpoint the exact location within a relatively large area where the meeting is believed to be taking place. There are more places satellites can't see than you might imagine, and Al Qaeda probably knows more about satellite limitations than a lot of us. What no country has is all the resources it would take to capture and analyze a single set of points on a digital earth. It is like trying to map the globe the way Google Earth has done over many years—only doing it every day. Then we have to keep doing it every day for years. In all that mass of data, we have to identify a gathering that is not a wedding party or other civilian social event, but rather a meeting of high-level Al Qaeda leaders.

When we finally come to the realization that more than one Al Qaeda member is in the same place, we could organize an airstrike, but who are all those other people standing around at the meeting place? They might be local politicians, a peace-making group, or merchants trying to sell them something. The United States, Israel and most Western powers go to great lengths to minimize collateral damage. This causes delays in taking out the bad guys and sometimes prevents attacks altogether. Other countries are willing to shoot first and assess later.

We need authority to strike, which normally doesn't take very long on something we are sure of. And we need a drone on the target location. Generally the drone has to fly to the target location, and by the time the drone is in position the target may have moved or left the scene. All this "seamlessly integrated" advertising is harder to put into practice in the world outside the Pentagon or a defense contractor marketing facility.

In April 2014 drones started firing into areas around the place where the press said the Al Qaeda gathering took place, giving us reason to suspect it actually happened where the reports said it did.[10] Yemen's government says as many as 30 Al Qaeda leaders were killed, but the U.S. press says 10–15, so it depends upon which story we believe. It is complicated and dangerous work, but that doesn't seem to prevent the airstrikes from being successful.

8. The Evolution of Command and Control Warfare 109

Hezbollah has been using drones since 2012 to do reconnaissance in Israel and has vowed to use them to deliver explosives.[11] Almost all of Hezbollah's drones are supplied by Iran, while Israel develops its own. Iran has one it calls the *Ambassador of Death*, which supposedly carries four cruise missiles and two bombs. They keep those close to home.

Drone strikes are to terrorists what Improvised Explosive Devices are to soldiers in the field. Both kill, but neither is war. Even bombing noncombatant public conveyances does not qualify as war in the world of public opinion. Strange as it may seem, the sinking of a single British passenger ship, the RMS *Lusitania*, helped draw the United States into the First World War similar to the way 9/11 brought us to war with Al Qaeda. The Germans sank the *Lusitania*. Someone pays Al Qaeda or ISIL, and that enemy has yet to be identified. It should be easy enough, if we apply ourselves to it.

9

PSYCHOLOGICAL WARFARE

In information war, psychological warfare is the use of information to affect the perceptions, intentions, and orientations of others. A person with an Internet connection, a tablet, two televisions and a smartphone might think this was the cornerstone of cyberwar, essential to winning the will of our target, but that would not be the case. Not yet, anyway. These things only work when the audience can be reached by one of those media.

The Russians have gone outside of the Ukraine to make a point about how they want the Germans to behave. Frank-Walter Steinmeier lost his composure over criticism (from both sides) that Germany has either too much sympathy and understanding for Russia or not enough. Demonstrators have picketed the speeches of German leaders, protesting "Nazi" involvement there.[1] The Nazi theme was the same as those billboards in Crimea and press stories in Russian newspapers. A consistent theme, applied through different media, is required to reach a number of different audiences, and not all of those are in Russia.

We are prone to believe that everyone and everything in the world has an Internet connection, and that the audience waits, with bated breath, for something to read. Anyone who uses the Internet doubts the wisdom of either proposition. When I recently cleaned out my mother's e-mail, she had 947 unread messages, some of them mine. That can hurt, but it proves the point that the Internet is no more effective than any other medium at getting people to read the messages they are given.

Only about one third of people in the world are connected to the Internet, with some of them connected in multiple ways, both mobile and fixed.[2] In most neighborhoods where this book is sold, more than that will have access. Some of the highest proportions of Internet access are found in Europe. The EU has almost as many Internet users as China, though China's usage is growing much faster.[3] Telephones and televisions are enough for some people—actually about half of the people in the world.

In the technology-centric cultures, we tend to believe the rest of the world still uses stone axes. There is a story about photojournalism that contains a beautiful picture, taken by freelance photojournalist Moises Saman, of the inside of a home in Helmand Province, Afghanistan (a province that has seen some of the heaviest fighting in the war).[4] Sitting on the floor, around an elder reading to them, were the other white-bearded elders of the village—some looking bored, some listening. It was cramped, and maybe the meeting had been going on too long. A picture of Hamid Karzai was propped up on the desk. This is how we think of Afghanistan: a country of villages, largely self-governed, run by old men in white beards.

Sitting in front of the senior elder was a cell phone, a symbol of connectivity to the rest of the world. There is no reason for us to see this as unusual, and yet several people who were shown this photo did find it at right angles to what they believed about the telecommunications of people in Afghanistan. The number of cell phones there may be surprising—18 million (more than Greece, Belgium, Sweden, and Israel, among others). There are over a million Afghan Internet users.[5] Yet few of us would think of Afghanistan as a "connected" country worried about cell-phone minutes and bandwidth in the same way the developed world does.

What is deceptive about these raw numbers is that the percentage of use in a country of 31 million people is not very high. They are largely illiterate. It would be easy to control the Internet use in Afghanistan, compared to someplace like China, where there are over a billion cell phones and nearly 650 million Internet users. If we wanted to control communications, we would want to live in a place that was small. China has to think bigger.

There is a difference in the way a country approaches influencing the will of people who don't have an Internet or cell-phone connection; they have to be reached in other ways. More than twice as many people have televisions, so the Internet is not essential to them. Neither is literacy. They may be watching the few television channels they can get with a TV antenna, but at least they have something. If we are trying to reach them, TV may be the best way.

An army private in a psychological warfare unit once told me what he was doing in Bosnia: "Playing music. Hour after hour, I put in tapes of music and sent those out over the radio." That was his perspective, because he wasn't preparing those tapes, nor measuring their impact on the population. Radio is just as good as television in places where TV means fixed-point transmission to boxes in people's homes. His tapes could be heard on mobile radios. They contained messages disguised as news for people listening to the music. This would be news slanted in favor of the position of the United States in solving a sticky mess that was a war between Christians and Muslims. It wasn't Tokyo Rose, but it was the same principle. This kind of verbal game is psychological warfare.

In some countries, literacy becomes a factor for large parts of the populations.[6] Could a person using a computer or watching television get a message that someone was trying to deliver if they were illiterate? They could see an image or listen to a soundtrack, and get the message without being able to read.

Literacy rates are as much political statements as factual representations of who can actually read and write in a given country. North Korea claims 100 percent of its people are literate, the United States 99 percent, Russia 98 percent, and China 95 percent. Even Syria claims 84 percent literacy, Iran 87 percent, and Libya 90 percent. Egypt has 74 percent, and this is an arbitrary line where we might consider that 25 percent are not literate; to reach these people, we need another medium. Countries like India (at 64 percent), Pakistan (55 percent), and Afghanistan (28 percent) have sizeable populations that cannot be reached by a normal computer exchange. Go back to that village in Helmand Province and think about those leaders having a 28 percent literacy rate, or soldiers we try to train to work together. Illiteracy complicates their ability to receive and give messages that can be understood. But those countries still have large TV audiences.

In a study by Harris Interactive and Northwestern University that looked at Middle East use of television and Internet, Egypt had much higher rates of television usage than the other countries studied.[7] Considering the literacy rate, that makes perfect sense. UAE, Qatar, Bahrain and Saudi Arabia had high rates of Internet use, and high literacy rates, but Egypt was much lower on both counts. Users in Qatar, Bahrain, Lebanon and UAE accessed the Internet in English rather than Arabic. There is no reason given for this finding. Governments trying to target messages to these populations use multiple languages and multiple media to get them across.

Egypt's television claims a part in the removal of Mohammad Morsi from power, which skeptics should examine in the light of television as a news source in Egypt.[8] In June 2013, six news directors discussed a common problem: their criticism of Morsi seemed to be followed by Islamists who surrounded the stations and intimidated their reporters as they came and went. They were saying, "We will kill you," which not very many people find difficult to interpret. To emphasize their point, they threw Molotov cocktails at the gates of the compounds, giving those threats more credibility.

What the news directors decided to do was both simple and straightforward: They started referring to the Morsi supporters as "terrorists." They did it every day, and they were consistent across the channels. We should see this in the same light as the Russian characterization of Ukrainian forces as "Nazis." It is a simple, powerful thing to do. A simple message, repeated often, is the foundation of persuasion. Advertising brings that home every day, focusing on the media most often used by the targets.

We tend to believe that any method of psychological manipulation is bad, but we should think about that idea more. One day, I happened by a yard sale and saw an interesting book for sale. It looked like a comic book and had Superman on the cover, complete with the DC logo. Old Superman comics are worth quite a bit, so I bought it and brought it home.

It turned out to not be in English, something I should have noticed before purchase, and it was no language I knew. Even so, it had a message that was clear enough for a child to understand. Superman was telling two young boys not to pick up those pieces that looked like small rockets. He flew them around to a village where someone had done that and blown up their house. They rescued a dog that was injured in the blast and took him home. The story ended with the two boys telling one of their friends about the danger of picking up these land mines, and Superman flew away happy. UNICEF had saved the day. This is probably the best use of psychological manipulation ever done. With this comic, UNICEF was trying to keep children from dying as a result of picking up small, anti-personnel mines.

Psychological warfare is better than that comic book. It has to be culturally sensitive, be able to be received by the target population, and be repeated (usually with different media) over time.

The billboards in Crimea were made for an audience of people, two-thirds of whom do not have Internet access. The Internet is not as effective there.

Influence by Internet

What the Internet is may not be as important as how effective it could be in influencing populations of people at war. There are two aspects—one intuitive, and the other not.

The first is the inability of an audience to concentrate on a message longer than Twitter allows. My speech professor used to say we would find it hard to keep a person's attention for longer than 20 minutes, and he usually proved to be right. E-mail and Twitter have been working on that number, bringing it down.

People coming into the workplace today are used to short messages (an e-mail is short by some definitions, but longer than Twitter) trying to influence them. Those messages are generally of high quality and entertaining, and they provide the reader with a structure to fill in blank spots that naturally occur. A snippet of truth here and there can do wonders. Workers make up stories to fit what is left out.[9] We could also say these new folks do not read well, can't spell, and use the Internet too often instead of thinking. Internet communication comes naturally to them. But the second aspect is the real reason most businesses find the Internet to be a difficult medium for effective persuasion.

The Internet is more than just moving images on a box, carrying more information than TV ever did. But, as Marshall McLuhan said about TV, the impact of the Internet and the cable providers is greater than the sum of their parts. Internet persuasion is no better than television persuasion when trying to get people to do something, but TV combined with the Internet has more power of persuasion. Where it may be better is reinforcing ideas that have already formed among like-minded people.

When Egyptian President Mohamed Morsi spoke to a crowd about his willingness to shed his own blood in the cause of staying in office, parts of that speech were in English. The Egyptian opposition forces displayed a green-colored laser across a building that said, also in English, *Game Over*. The same slogan appeared in the same language at a rally in Kiev. Neither group was appealing to the majority of their countrymen, especially that 25 percent of Egyptians who are illiterate. They were instead trying to reach people outside their own countries.

For years, I taught a course on security education that reminded students of how hard it is to sell an idea to anyone, or, more importantly, to get workers to act the way someone wants. We wanted others to know that computer security procedures were important and what actions they should take, thinking these "facts" would change the way the people performed. But, like advertisers, we found that beliefs and performance are two different things.

When we see a television commercial for Michelin tires, we must know that there is an implication from the qualities of tires we see in that ad that we should buy Michelin tires, if we need tires. Advertisers expose viewers to that ad as often, and in as many different types of media and languages, as they can. They use magazine ads, television, radio, billboards, and the like. They do that because most readers do not get information from the same sources. Some like newspapers; some like online sources; some like radio. One medium is not enough.

What determines how effective advertisers are is whether people who see the ad actually buy Michelin, and not some other brand. They are not trying to make the audience *like* Michelin; they are trying to get them to *buy* Michelin. The Russians in Crimea would not have found it satisfactory for their audience to like Russia; they needed to stay out of the way of the takeover, or actively help in carrying it out. From that standpoint, the Russian campaign was successful in Crimea, but may not be doing as well in the eastern part of the Ukraine.

Michelin knows whether the amount it spends on advertising is balanced by an increase in tire purchases. If it didn't give the company good results, it would not take long to stop. If UNICEF found that those comic books distributed among the population did not result in a decrease of children picking up anti-personnel mines, they might want to stop making the books. We have to wonder if anyone looks at the success or failure of programs like this.

There are four basic principles here: people don't get their information from a single media source; the sources are mostly voluntarily initiated; the goal of any campaign is to move people to action; persuasion is not influenced by "facts."

The U.S. Secret Service recently ran an operation against a Russian hacker that raised some concerns in Moscow, and sounds more like a spy story than the badge-carrying Secret Service.[10] The individual arrested, Roman Seleznyov, came to the United States by an unusual route—from the Maldives, an archipelago off the southwest coast of India, famous for its beaches, to the U.S. territory in Guam.

The Russians, who are good at this type of manipulation, have great respect for his father, Valery Seleznyov, Member of Parliament. They have repeated the MP's unlikely suspicion that his son was kidnapped to exchange for Edward Snowden. It may not be true, but it gets readership of their main issues. On the same day this story broke in the United States, the official news service, Russia Today, ran the story, along with others having similar titles, to show their "concern" for Roman Seleznyov in an indirect way:

- "Foreign Ministry Concerned over U.S. 'Hunt' for Russian Citizens in Foreign Countries"
- "Moscow Rips into 'Vicious Practice' of Extraditing Russian Nationals to U.S."
- "Russian Official Slams U.S. for Turning Down Moscow's Extradition Requests"

ITAR/TASS said Seleznyov confirmed detention of his son. In addition, United States Attorney Jenny A. Durkan, Western District of Washington, issued a statement saying that "a Russian man who was indicted in the Western District of Washington for hacking into point of sale systems at retailers throughout the United States was arrested this weekend and transported to Guam for an initial appearance."

"Roman Valerevich Seleznev [sic], 30, of Moscow, also known as 'Track2' in the criminal credit/debit carding underground, was indicted in March 2011 for operating several carding forums that engaged in the distribution of stolen credit card information," the statement went on to say. "At his first appearance in Guam today, Seleznev was ordered detained pending a further hearing scheduled for July 22, 2014," where he pleaded not guilty. He was alleged to have stolen credit card numbers from point of sale systems for two years, starting in October 2009. Seleznyov was also charged in an indictment filed in the District of Nevada that was unsealed on November 13, 2013, alleging that he participated in a racketeer-influenced corrupt organization, conspired to engage in a racketeer-influenced corrupt organization, and possessed counterfeit access devices.[11]

The slant for the Russian stories shifts the emphasis to how the individual was brought to court, rather than why, possibly because these credit card rings are widely ignored in Russia, and the involvement of the son of such a high-ranking government official makes an embarrassing arrest worse. This approach deflects the story to focus on the U.S. actions rather than Seleznyov's behavior.

10

Follow the Money

Modern war is a level above what we see and hear about in the news. If we follow the money for what we call war these days, we will not have too much trouble figuring out who funds that part of conflict. The ones who do so are not the makers of cyberwar. They will get there one day, but for now they fight kinetic wars because that is all they know.

What separates parts of the world from the United States is control of information, both to and from populations. Other countries have trouble with it, but they are getting better at controlling every aspect of information flow within their borders (and a good deal of it outside). Conversely, we make a conscious effort to not control ours. We strive to maintain a free press, but some countries want to control theirs. Internal spying on their own citizens is prominent. They use information control to wage war with us, in ways that have never been used before, trying hard to not become an enemy in the process.

Afghanistan, which is not receptive to any guidance from its liberators, has had war that has gone on for decades, and ignores almost everyone with a hand in winning or losing. The Afghans have taken to thumbing their noses at the allied forces, and the Russians and British before them, as they wind down to a future withdrawal of the allied forces. While the allies have tried to impose democracy, the loser of the first major election has decided to form his own government. This somehow makes sense in Afghanistan. Some people think they are ungrateful, but we might want to think about that idea longer.

We assume, without really thinking about it, that armed conflict is between the people who are protagonists in a fight, even though we know better. The Afghans are fighting somebody else's war, and getting paid to do it. So is Al Qaeda; so are the Islamists in northern Mali; so are the Somalis; so are the Syrian fighters on both sides; and so is ISIL. They are closer to mercenaries than national warriors. We define those serving in the Taliban as enemy insurgents, but they are mercenaries. By the same token, we define those working for the Afghan army as allied soldiers.

President Carter signed a finding in the 1970s to have the CIA intervene in Afghanistan. The Russians were concerned that the CIA was involved there (although it is uncertain whether they knew about Carter's finding, there is no question that the KGB knew the Americans were there). The CIA also knew the Russians were there, since several of their military advisors had been executed in Herat a few months earlier.[1] When the Russians invaded Afghanistan, the Iranians started doing the same things to them that they had done to the United States—attacking their embassy, harassing their business interests, and shutting off supplies of gas (which Russia didn't need anyway). Iran and Russia didn't get to be fast friends until the Russians left Afghanistan, never to return. This is the way of friendships in the Middle East. They have ups and downs.

The Russians are involved in Chechnya, and that is not going much better than Afghanistan. The rebels had trouble getting enough money to operate. In better times, they relied on part of Chechnya's business community, as well as Islamic charities in the Middle East, Jordan and, strangely enough, the United States, to survive.[2] The Russians seem to have been able to choke off some of these resources and have forced the terrorists into lower-cost operations, like kidnapping and taking schoolchildren hostage. These kinds of things don't endear the rebels to anyone in the civilized world and make getting money even harder. ISIL should have figured that out by now, but seemingly hasn't. Most of the countries of the West do the same things to Al Qaeda, with varied success.

The point of this financial clampdown is a recognition that money keeps these operations going, and cutting off the flow of money helps to slow things down. But money, like operations to expand their influence, is covertly given. It is harder to choke off the supplies when we don't know who is really giving money to someone (although the Russians seem to have figured out a way in Chechnya).

It is difficult to make the Saudis our enemies, any way you look at it, and the charities are splintered into groups of people trying to help those Syrian refugees in Turkey without spilling any of that loose change to the Kurdistan Workers Party (PKK), which the British Foreign Office warns travelers about, even though it fights with the Iraq government against ISIL. That has come together in the town of Kobani on the Turkish border. Turkey is not going to help the PKK there, and ISIL makes progress every day toward taking over the town. The British warnings about the PKK say its members kidnap people and blow up public areas into which tourists may wander. It doesn't appear that the PKK gets included in the Islamic charity handouts, but nobody knows for sure.

In Syria, Assad blames "mercenaries with connections to Al Qaeda" for attacking his government, a point that has some truth. At the same time, Assad hires mercenaries to augment his own army and the Iranians hire

Afghans to fight alongside those.³ They are all hired guns to somebody. The trouble with any group of hired guns is loyalty, which lasts as long as the money to pay them. Sometimes, even that isn't enough if somebody else will pay them more.

We have a good example in Cheick Aoussa, a former poet who worked as a mercenary for Moammar Gadhafi, who can no longer pay him, because he is dead.⁴ The French, who have always seemed to be the first to fight a war that nobody can win, have decided to try to make peace between Aoussa's Islamist friends and the rest of Mali. Aoussa spent a good part of his year in hiding, in the Sahara Desert, a place that seems unlikely to provide much cover for anyone. The *Wall Street Journal* said he had rejoined his tribal friends (the Tuaregs). A search for them in Google Images will uncover pictures that make them look exactly like the army Peter O'Toole put together in *Lawrence of Arabia*.

The little village of Kidal looks like any place in Afghanistan. There are plenty of guns, Islamic militants, and government military opposition, just over the horizon. From here, Aoussa told Al Jazeera that he had changed the name of his organization to distance himself from the claim that he supported Al Qaeda.⁵ He wanted to be known as a Mali patriot, who just happened to favor things that Al Qaeda seemed to fight for. He would fight the government's army if it came, he said, but he did not mention what would happen if the French came with them. Kidal appears to be the last bastion of people like him in that part of the country. There are hundreds of little villages like Kidal and Kobani in the world.

Most of us can see that this is manipulation of the press to meet a political objective (and we thought only candidates for public office were good at it). Aoussa is not going to get much money from the supporters of Arab causes if he aligns himself exclusively with Al Qaeda, since two-thirds of the countries of the world are trying to find out where Al Qaeda gets its money and shut off the flow. The other third is filling the financial pipeline. This can make for an erratic funding source. If Aoussa can convince people with money to support a patriot, he can broaden his support.

Afghanistan is much the same. It is more than the sum of villages and towns that have repeatedly been fought over. There are geographical clans of people with their own standing militias and private guards, and they all depend on war. There are government officials, police, rescue workers, hospitals and service workers who likewise depend on one war or another for their living. Of the $6.7 billion we used to spend monthly on the war in Afghanistan, a percentage went to local labor for various humanitarian programs, fuel, transportation, schools, and other types of support. Part of it may even have gone to the Taliban. The Afghans are good at perpetuating profitable wars, something they have done for 200 years.

One day, the countries of the world will get together and say they aren't going to fund this anymore, but history says that will be a long time from now. The Afghans are part of the business of war, a checkerboard of sovereign nations, mercenaries, and arms dealers, all struggling for a little bit of power and a market for their products.

In Mali, Afghanistan, Libya, Lebanon, Syria, and many other places, there is no national army defeating a foreign invasion; it is instead some groups banding together to defeat one, and sometimes fighting each other in the process. Each group has its own armed militia, and they fight because someone, who keeps them employed, wants them to. They have been doing it for a long time because the Afghans and people like Cheick Aoussa make a living from fighting wars. This further confuses the issue of whether they are at war with anyone.

When we refer to the war in Afghanistan or Mali, we are not saying either country is at war with anyone, but somebody is at war there. By focusing our attention on the fighting, instead of the funding, we miss the aspect of global war. It seems like nobody is at war, when in fact quite a few of them are.

So, we might ask ourselves, who is benefiting from having terrorists beat up on the rest of the world? The Russians have had their share of it with the operations in Chechnya, so they don't get much benefit from having terrorism get a boost from anyone. The Chinese have the Uighurs, who regularly stir up trouble in the northwest of their country. The northwest parts of China and Chechnya are both majority Muslim populations in countries that aren't. China claims only 1.8 percent of its people are Muslim, in a country where only half officially have any religious affiliation; of these, one fifth are Buddhists. Russia is mostly Christian.[6]

In March 2014, 29 people were killed in China in a bizarre train station stabbing, where commuters were picked out at random, and then calmly killed; some of them were children.[7] One of the perpetrators of a similar ax-swinging incident in a game center confessed on television that a friend had convinced him to join in a holy war.[8] In a third incident in Guangzhou, six people were injured by knife-wielding terrorists.[9] In a shopping area in the same part of the country, 31 were killed when two cars packed with homemade explosives caused a bad day for the middle class.[10] The Chinese blamed the Uighurs for these problems, and several similar incidents, and cracked down on them with military force.

The Chinese have good reason to believe websites that promote jihad as the right path against China may be helping extremists in their pursuits. These sites offer instruction in suitcase-bomb making and other types of terrorism that China would not like to have happen.[11] The Uighurs say they will get revenge, and post it on their websites. To the UN, these are warring factions, but this is war of a different kind.

10. Follow the Money

China has tried to stop the Uighurs by appealing to those who train them—Turkey and Pakistan. It even went to Pakistan's Inter-Service Intelligence (ISI) to appeal to them directly because it believed "freedom fighters" were being trained there.[12] As many as a thousand of these fighters were said to belong to Al Qaeda, and a hundred of those were fighting the allies. Some were even recruited into the Taliban. China's neighbors—Kazakhstan, Kyrgyzstan, and Russia—used to allow immigration of disaffected migrants leaving China. Now they are not, leaving them bottled up in their own region, Xinjiang, causing more trouble for China.

If we were looking for countries that benefited from this kind of situation, we might think that nobody does. If that were true, there would be no terrorists anywhere, because nobody would pay to keep them going. Somebody does.

Aside from a very few authoritative sources, there is not much written officially about terrorist financing, but state-sponsored terrorism points back to Iran, Pakistan and Saudi Arabia.[13] War works for those who finance these operations. China sees it the same way, leading the Chinese to go to Pakistan to try to kick out Islamic militants training the Uighurs.[14] We wonder if the Taliban is part of Pakistan's army, while the Taliban also fights with Pakistan, to keep the Pakistani leaders from getting their country back under control. It is unclear which side anyone is on.

When it comes to finding out who is funding the terrorists, nothing has been more entertaining than civil suits brought by victims. If my son was blown up on a bus riding around in Israel, I'm not sure a lawsuit would be on my list of things to do to get at the guilty parties, but, as it turns out, it couldn't hurt. Lawyers are very tenacious advocates for the victims, though they have not been very effective at getting the cases into court, or even settling. What they have been good at is stirring up governments. Getting someone from the Israeli Mossad to testify in open court is about as likely as getting someone from the FSB or CIA. It doesn't happen very often.

In a case notable for its postponements and delays, one such agent, Uzi Shaya, may be allowed to lay out a case in open court that accuses the Bank of China of knowingly allowing one of its branches to maintain accounts for terrorist organizations. Shaya is expected to testify that he met with Bank of China officials in April 2005 to warn them that Hamas and the Palestine Islamic Jihad were using accounts to launder money and finance attacks on civilians in Israel. China has denied receiving any such warnings, and brought pressure on Israel to drop its support for the testimony. The case filing clearly shows that Israel tried to stop China from maintaining the funds and, failing, fed information "to private lawyers, with the express intention of prompting American victims to sue the bank."[15]

In a second case, the Arab Bank PLC and two other organizations, the Saudi Committee for the Support of the Intifada al Quds and Al-Shahid Foundation, were accused of helping fund terrorist activities by giving money to Hamas and the families of suicide bombers, as well as several organizations on Israel's blacklist of Unlawful Associations.[16] Where most legal battles have two protagonists, this one has more. The same law firm, with different plaintiffs, is bringing action against Credit Lyonnais, S.A., for the same crime. Nobody is saying Israel gave this firm anything.

In the first case to be decided, a jury, after only two days of deliberation, ruled that Arab Bank PLC (which had successfully fought a similar suit a few years earlier, and agreed to make changes in its operations without admitting guilt) was providing funding from Saudi donors to Hamas. It was the first civil action in a terror-financing case to come to trial in the United States.[17] Records showed transactions through the bank's New York branch, and the case was filed under the Anti-Terrorism Act of the United States. On the Arab Bank's side of things, a statement was posted on its website:

> The Saudi Committee is an internationally recognized humanitarian organization that was established to raise money for humanitarian aid and programs that would support the Palestinian people who were injured during the Second Intifada. Specifically, the Saudi Committee's funding was designated to support the unemployed, those injured or hospitalized because of acts of violence, those whose houses were destroyed, and to support Palestinian schools, hospitals and infrastructure. While plaintiffs have alleged the Saudi Committee was engaged in supporting terrorism the organization has never been designated by the U.S. government on the Office of Foreign Assets Control list of prohibited parties.

The other side of the story is also on the website of the Osen law firm that represented the claimants.[18] Osen states that the Israeli government issued a report critical of the Saudi Committee for the Support of the Intifada al Quds,[19] indicating the Arab Bank's Gaza branch maintained accounts for families of suicide bombers, using money supplied by the committee. There are quotes from the Chairman of the Board of the Bank citing what is certainly a lack of enthusiasm for Israel. The claimants have also listed several "terrorist organizations" that maintained accounts with the bank, and in the coming legal proceedings, they will be trying to show the bank knew that it was supporting terrorists. They cite a Treasury Department designation of these five organizations as Specially Designated Global Terrorists[20]:

(1) Commite de Bienfaisance et de Secours aux Palestiniens (CBSP), of France.
(2) The Association de Secours Palestinien (ASP), of Switzerland (an organization related to CBSP).
(3) The Palestinian Relief and Development Fund, or Interpal, headquartered in the United Kingdom.
(4) The Palestinian Association in Austria (PVOE).
(5) The Sanabil Association for Relief and Development, based in Lebanon.

10. Follow the Money

We certainly know that most of the money was given in good faith, but deciding the bank's knowledge carries over to some of the other instances of support for groups less friendly.

In the same way, somebody pays the Taliban to operate. A small part of its funds is said to come from "taxes" on contracts the allies pay for in Afghanistan.[21] We know that half of the Taliban's money comes from cocaine, and Afghanistan is the number one grower of poppies in the world. Taliban agents are in a perfect spot to make that work for them. Part of their money comes from Islamic charities in Persian Gulf states. What the victims' cases have shown is the tangle of those charities and the difficulty of proving which ones are directly supporting terrorist acts.

The Intelligence Service in Pakistan (ISI) is often associated with Taliban support. A UN report, collecting interrogations of 27,000 people, says the Taliban gets direct support from ISI, and that ISI knows where Taliban leaders are hiding. Pakistan denies any such arrangement.[22] It has even launched attacks on the most inaccessible terrorist enclaves, in Waziristan, a dangerous and protected place in the northwest. It is hard to see which side Pakistan is on, which may be the intent.

We know who pays the Afghan army.

Somebody pays the FARC, Boko Haram (lately known for kidnapping young girls), Ansaru, Al Qaeda in the Arabian Peninsula, Abdullah Azzam Brigades, Jemaah Islamiyah, and a hundred other terrorist groups, criminal gangs, hackers, and those who have staked out territory in the dark world of mercenaries. These are the people paid to fight our wars. Countries we think of as both "allies" and "not-allies" use mercenaries to fight their battles, without ever calling them by that name. When Syria started fighting internally, they came from everywhere to engage.

According to a number of accounts, Irish, Russian, Saudi, Kuwaiti, Moroccan, Lebanese, Israeli, Greek, Libyan, Pakistani, Afghan, Turkish and Iranian fighters have come to fight in Syria. There are about 7,500 foreign fighters, coming from 50 countries.[23] These stories are not all true, but the full truth of any of them will never be known. It is hard to get an accurate body count in the Syrian war, and even harder to check passports of the dead. We should say, however, that a good number of mercenaries are fighting on both sides, and somebody is paying them to do so. They don't work for free, but the King of Jordan told CBS's *60 Minutes* that the group made $6 million a day from oil sales, and was largely self-financed. The United States and its allies started bombing those oil wells shortly thereafter.

The Treasury Department indicated in a 2011 report that Al Qaeda and Iran had a business arrangement that allowed them to have some operatives in Iran and to facilitate operations in Afghanistan, Qatar and Pakistan. Money was provided to Al Qaeda by operations in these countries. Under the agreement, prisoners in Iran were released and sent to Pakistan to resettle. (We would have to ask the Pakistan government if it was aware these relocations were taking place.) The report makes the following conclusion:

> "Iran is the leading state sponsor of terrorism in the world today. By exposing Iran's secret deal with al-Qa'ida[24] allowing it to funnel funds and operatives through its territory, we are illuminating yet another aspect of Iran's unmatched support for terrorism," said Under Secretary for Terrorism and Financial Intelligence David S. Cohen. "Today's action also seeks to disrupt this key network and deny al-Qa'ida's senior leadership much-needed support."[25]

In the post-9/11 clean-up of investigative strings, Congress wondered why terrorist financing did not get more attention from the government organizations responsible for keeping an eye on terrorism.

In Lee Hamilton's testimony, there were reasons enough: the FBI investigated but did nothing to pursue those involved; the justice system did nothing in the way of prosecution; the intelligence community had little understanding of how some of these financing operations actually worked; the mixing of legitimate funding and Al Qaeda funding was difficult to detect and act on. On the matter of Saudi Arabia, Hamilton said:

> The conclusion was that we found no evidence, as you have stated correctly, that the Saudi Government as an institution or as individual senior officials of the Saudi Government supported al Qaeda. Now, we sent investigators to Saudi Arabia. We reviewed all kinds of information and documents with regard to that that are available in the intelligence community. We listened to many, many people who talked to us about these things. We followed every lead that we could.... We did find in this, the pre-attack period, pre–Saudi Arabia attack period, that there was a real failure to conduct oversight in the Saudi Government, there was a lack of awareness of the problem, and a lot of financing activity we think flourished. We think that Saudi cooperation was ambivalent and selective, and we were not entirely pleased with it.[26]

That hasn't changed very much since 9/11. In an interview with Radio Free Europe, Michael Jacobson, who served as a senior adviser to the U.S. Treasury Department's Office of Terrorism and Financial Intelligence and as counsel on the 9/11 Commission, said the money still flows from "private sources" in Saudi Arabia, Kuwait, Qatar, and state sources in Iran and Syria.[27] A former Treasury official recently said the same of ISIL, fighting in Iraq.[28]

In the same way, somebody is at war with Israel, and we think it is easy to tell who that would be. A succession of presidents has tried to produce an agreement between the Palestinians and the Israelis, because there are constant battles between them. If we start looking for the source of the trouble, however, we could see disagreement, maybe even some confusion, about the culprit. Hamas largely governs Gaza, where most of these people live. Hamas also ends up with most of the money given to the people of Gaza in the name of humanitarian aid. The Palestinian Authority receives money from the European Union, the United Nations, the United States, Saudi Arabia, and other Arab League countries. This money does not seem to help the Palestinians lead a better life.[29] Therefore, it must be going for other things. We know some of it went into tunnel linings.

These days fighting is the norm, and somebody wants it; somebody pays for it, mostly under the table. We can lay a substantial amount to religious charities—Christian, Jewish and Muslim. All of them claim they offer aid for humanitarian reasons, though they sometimes have to split hairs to make that characterization. This is not about arms shipments, because none of these groups say they want to ship arms for humanitarian purposes. They are all trying to help out people who face followers of other religions in showdowns that are not representative of the path any of them says they follow. In March 2014, ABC News showed the faces of CAR Christians who had been attacked by Muslims. They were angry and swinging machetes like ancient swords. Retribution, revenge, and self-defense are all words being applied to the various groups. Oddly, these are not characteristics that religious groups like to attribute to themselves.

Maybe Lenin was right. He quoted von Clausewitz, saying war is an extension of politics, but he added that war was a perversion of it. He called this *revolutionary defencism*, which is an excuse for war that leaves the same people in charge after the fighting is over.[30] Lenin thought that was a bad idea, and went on to implement a new approach in Russia.

China and Russia are Iran's best friends, so they probably know what Iran has been up to, and tolerate it. They should think about how that plays out. We play the same game with Israel and Saudi Arabia.

In 2001, the French ambassador to London said that "shitty little country, Israel," caused most of the trouble in the world.[31] Different people have different perceptions about who causes trouble, and a good many of them include the United States on this list. We seem to be confused about this very simple concept, but we shouldn't be. Everyone knows where the Israelis get their money.

Since World War II, the United States has given Israel $118 billion in aid, almost all of it military assistance. By agreement with the Bush and Obama administrations, Israel receives a little over $3 billion a year in foreign aid, specific military aid for missile defense systems, and benefits no other country enjoys.[32] In addition, Jewish charities in the United States give almost a billion dollars a year to Israeli causes.[33] While several hostile leaders have threatened to "wipe Israel off the map," those leaders don't seem to be very effective at doing it. Most of them have even stopped saying it out loud.

Now and again, we do ask why we give Israel so much money. It isn't just demographics, because the United States has less than six million Jews, and almost four million Buddhists. In addition, there are four times as many unaffiliated people in the United States than Jews and Buddhists combined.[34]

Israel certainly doesn't act like a mercenary military force of the United States, although we might say the same thing about Afghanistan, Turkey, Pakistan, Egypt, Jordan, Colombia, Somalia, and Russia. We give almost 1 percent of our military aid to all of these countries, with Afghanistan taking nearly half of that amount, and Israel a fifth.[35] Has anyone asked why we needed to give the Russians $127M in military aid?

Israel experiences attacks on its soil almost every day, most of them from Gaza. We can count the days until the next one; there won't be many. In March 2014, Israel was attacked by forty-one short-range missiles, five of which fell on populated areas in the south. In July, 1,500 missiles (or more) were launched in eleven days, and they were showing a longer range than those from March. Israel responded each time by attacking "terrorist training sites" or missile assembly areas in Gaza.[36] As we look back on those days, they seem relatively peaceful compared to what came next.

Tony Blair, the special ambassador to the Middle East and former British PM, says the problems of Hamas and Israel are rooted in the place of religion in politics. (He would know, since he spent six years being frustrated by a lack of results in his diplomacy.) In the area, Hamas can't make up its mind what that place should be.[37] The will of the people is to fight, even when it isn't good for either side. The Islam enterprise has always shown strains between the Saudis and Iranians, which are closer to the roots of the disagreement.

Israel is equally known for its prevention of attacks, especially targeted assassinations of Hamas leadership and interdiction of weapons shipments. Until 2013, the Israelis did not admit that they attacked Hezbollah, and both the Syrians and Hezbollah did not respond to attacks they thought might come from Israel. Hezbollah is funded largely by Iran.[38]

The United States treats Hezbollah as a terrorist organization and has sanctioned a group of companies in Lebanon, the United Arab Emirates, and China for acquiring more sophisticated weapons for Hezbollah.[39] The main supplier is a company called Stars Group Holding, of Beirut, which had its assets frozen in the United States. The organization was accused of buying engines, communications equipment, electronics and navigation systems, and enhancing their use of drones for surveillance. In July 2014, Israel shot down one of them with a Patriot missile.[40]

We have to ask ourselves how this Hezbollah-Israel façade is maintained when the Israelis don't seem to mind admitting that they attack Hamas. They are even in a war with Hamas, invading its part of Gaza. In part, it is control of information about who is attacking whom, and who admits what. Hezbollah is a violent organization, but less is said about that than about Hamas. This leads to some inconsistent behavior.

In my first book, I used the example of the Stuxnet worm. When the story first broke, Iran denied anything had happened, saying it had no effect on production of plutonium. Israel denied having any part in the operation. The worm corrupted Iran's centrifuges working on nuclear material, which might have been construed as an act of war. Both countries denied it happened, and neither was at war with the other (at least on that occasion). In the same way, Hezbollah and Israel avoid being at war by denying or ignoring the attacks. Neither of them wants the consequences of having an announced war, but they still fight one.

The week before the March 2014 missile attack, the Israelis intercepted a ship in the Red Sea, boarded it and seized missiles made by Syria and headed for Gaza.[41] We saw video of this event on almost every news channel in the world. Israel blamed Iran for this shipment, which was important because these were longer-range missiles, meaning Iran was upping the ante in the war—but with no mention of Hezbollah.

Iran responded by posting a video on YouTube showing U.S. planes being blown out of the sky by Iranian ground defense missiles and what looked like a nuclear mushroom cloud over Israel. This may be counterproductive for the Iranians, since a nuclear weapon going off is not the outcome of a purely peaceful use of nuclear material. But the audience they are targeting is not living in Europe or the Western Hemisphere, where that would be a factor. The Western press seems to have missed both the symbolism and the posting on YouTube (http://www.youtube.com/watch?v=uY3w7yjasc4). The video is in Farsi, but it isn't necessary to speak the language to get the message from it: The Iranians clearly are worried about a strike on Iran's nuclear facilities, and seek deterrence, perhaps, with an image of war that only a nuclear weapon could bring.

There is almost nobody who thinks Iran is a friend of ours or, for that matter, of too many other countries. We tend to favor the Saudis and the Israelis. That's important because the Saudis and Iran have been at odds over religious matters for centuries. Neither of them favors Israel, but they are preoccupied with building power over a single, large, but fragmented religion. The factions that can claim to control Islam will run an empire. There are 1.6 billion Muslims in the world, and, contrary to popular belief, the majority does not live in the Middle East. More Muslims live in Pakistan and India than in all of North Africa and the Middle East.[42]

Thought leadership of such a large population is a goal that is difficult to achieve, but the cradles of Muslim thought are in Saudi Arabia. Muslims still return there for their annual pilgrimage, and it irritates the Iranians to no end. That is why an agreement with Iran over production of nuclear material has the Israelis and Saudis on the same side of a complicated issue—something that rarely happens.[43]

The United States, Britain, Russia, China, France and Germany have tacitly supported an agreement that allows Iran to continue to have some nuclear material. Within moments of the announcement in 2014, there was disagreement about what "uranium enrichment" actually meant, and no discussion of some of the other issues, like Iran's heavy-water reactor project, its suspected nuclear military research, and its ballistic missile program, which are all mentioned in the joint plan of action agreed on between the parties.[44] The agreement is thin, and it has not been well received by the Saudis. President Obama felt the heat on that in March 2014 when he visited.

We may have forgotten where Iran got most of its nuclear capability. Pakistan has probably done more for Iran to get a nuclear weapon than any other country. So, when we start to describe behaviors of potential enemies, giving nuclear technology to an enemy of ours might be something we should consider. Abdul Qadeer Khan, a prominent nuclear scientist, often accused of spreading his knowledge around the Middle East, says the Iranians tried to buy a nuclear weapon from Pakistan in the 1980s.[45]

As Khan's story goes, Pakistan declined to sell, but gave the Iranians bomb-related drawings, parts for centrifuges to purify uranium and a secret worldwide list of suppliers. "Iran's centrifuges, which are viewed as building blocks for a nuclear arsenal, are largely based on models and designs obtained from Pakistan," the article says. The Iranians say they are only using these centrifuges to build up nuclear material they can use to generate electricity. If that were the case, they wouldn't have needed a bomb, and Khan's story would be fiction.

Khan claims senior military officers in Pakistan, and possibly some political officials, knew about what he was doing and encouraged it. When the Iranians asked for a bomb, they were told they could have one, but only if they built it themselves. They certainly seem to be moving in that direction, assuming Khan's story is accurate.

More than that, Khan's list of customers included North Korea, Iraq (before the war), and Libya (before its revolution). He was selling technology and access to a privately constructed list of suppliers. These suppliers were not ones to ask questions about the intended use of the items they were selling. He put together firms in South Africa, Malaysia, and Turkey to help with manufacturing. Khan's story that the government was cooperating may not be true, in which case the nuclear technology was being transferred without their knowledge or consent. We should be equally concerned about Pakistan's ability to protect its own technology, as we might be that Pakistan cooperated in the transfers. Neither is a good thing.

10. Follow the Money

Saudi Arabia and Iran have a proxy war that is broader than the Middle East. Besides Hezbollah and Hamas, they support opposing sides in Syria, Iraq, Egypt, Pakistan, and Lebanon. The Saudis don't want Iran with a nuclear weapon, and in that respect they have most of the rest of the world in agreement, including Israel. Israel bombed a nuclear plant in Iraq in 1981, so the region is concerned that it might want to do the same thing again.

This disagreement has been the root of a large part of the world's current wars, but these are regional conflicts, not global issues. The protagonists have not figured out yet that leadership is going to depend on winning the will of the people in these conflicts, not killing off the opposition. They should be watching the Russians in the Ukraine.

11

A Military Left Out

Militaries, and their supporting contractors, have difficulty defining a role for themselves in modern war because much of it is outside their domain. In all but a few cases, military capability is not as relevant in war as it once was. In the previous examples, most aspects of information war, particularly the cyber components, are executed by people and machines that are not military.

It is one thing to be sitting at my desk at home and have my computer go down, and quite another to be lying in a hospital bed when that attack comes. If the computer on that little pole next to my bed goes down, it may stop pumping medication into my system. Most people do not see the connection between collateral damage on a battlefield, where combatants attack each other, and damage to a network that we rely on for safety or security, but for the little girl who gets run over at an uncontrolled intersection, or the person connected to a life-support machine, the results are the same. In this kind of cyberwar, the combatants aren't the ones who die.

The United States described its concern with the new Internet in Presidential Decision Directive 63, "Critical Infrastructure Protection." That was in 1998, which seems like an eternity ago. The main thrust is the impact to the economy by attacking telecommunications, energy, banking and finance, transportation, water systems, and emergency services, both governmental and private, either by sector or all at once. The military is responsible for only a small part of this infrastructure. When the directive was written in 1998, there were huge concerns with attacks made on these sectors, and on the military, which is not directly mentioned. There was a mistaken belief that the military could take care of itself, but the public sector couldn't. Where that idea came from, nobody knows.

Electricity is only one of the aspects of the critical infrastructure that are at risk. For example, air, road, truck and rail traffic are all managed by computers. In July 2012, the Washington, D.C., area got a taste of what that means when the computers that ran the Metro were down. These computers allow rail managers to see where the Metro cars are located and handle switch-

11. A Military Left Out

ing. Metro officials said they shut down the trains for 40 minutes "out of an abundance of caution," doing it again four days later. Metro North and Amtrak, in New York, were down for two hours when computers and backups failed, a repeat of another incident in September 2013.[1]

Just in the last few years, air traffic control systems have failed in Southern California at Los Angeles, Palmdale, John Wayne and San Diego, as well as in Kansas City, Great Britain, Ireland, Atlanta and intermittent locations on the East Coast, including one near Chicago that backed up flights for days. The FAA issued a statement in response to the Atlanta outage that describes some of the complexities of this type of system:

WASHINGTON—There has been a major, nationwide air traffic control computer system outage this morning, and it is having a severe, negative impact on air travel across the country. Here is what we know thus far:

—The NADIN system (National Airspace Data Interchange Network), which is the computerized system for processing flight plans and information for every flight in the country, has failed in both of its locations—Atlanta and Salt Lake City. We do not yet know the technical reason for the failure. We have some reports that the system is coming back online, which is good news, but there will be flight delays throughout the day due to the "ripple effect" of this outage.

—The NADIN failure has created a domino effect of problems throughout the country, starting with the inability of FAA automated ATC systems at major regional facilities to process flight information, forcing the manual input of information by air traffic personnel. Air traffic controllers are without electronic decision-making tools and cannot keep up with the sheer numbers of flights—resulting in delays.[2]

Without power, all these systems eventually fail, unless emergency generators can have fuel delivered at regular intervals. After that, people will die or be injured when accidents start to happen.

What we have seen in the last couple of years is an escalation of that capability. When groups go after the banking, police, fire and rescue, air traffic control, emergency telephone notification and energy sectors, they are slowly gaining control of large sections of the national infrastructure. These sectors are so large that the intent behind such actions has to be to cripple the countries thus assaulted.

At the basic level, the kinds of attacks we are seeing today are not going after a large part of what is considered the critical infrastructure of the United States. That may seem incongruous to some people, but that national infrastructure is a good bit larger than most countries can think of going after. Presidential Policy Directive 21 describes what is included: chemical facilities, dams, communications, commercial facilities, manufacturing capabilities, the defense industrial base, emergency services, energy, financial services, food and agriculture, government facilities, healthcare and public health, information technology, nuclear power, transportation systems, and water and wastewater systems. We could never argue that these are not important sectors of some countries' economies, but this broad approach is why we don't have much going on to protect them all. They are far too big in their scope.

When I served on the President's Critical Infrastructure Protection Committee, a big concern was getting priorities set for national initiatives. There are 13,500 chemical manufacturing facilities in the United States, 75,000 dams, and 100,000 facilities in the industrial base. There are also 100 licensed nuclear power plants, 140,000 miles of rail, 52 major airports, and so on—as you can see, there are quite a few things that Homeland Security thinks need to be protected. Besides that, there are bridges, government buildings like the White House, and electric distribution systems that vary from defensible to indefensible. There are too many targets.

I once went out to a power plant in Iowa where the power lines ran up a steep 40 percent grade to the top of a mountain before entering the first major town, and then I looked at two airports and some manufacturing centers in Illinois. My military background was in nuclear security, so I viewed anything like that power system or an airport as impossible to defend. They are too big geographically and have budgets that are too small to defend everything that keeps terrorists home at night. Defending everything means defending nothing well.

Homeland Security, which runs most of the protection efforts, has yet to figure out what prioritization means. Its leaders seem to believe that putting a little money into everything in their area of responsibility will make the world a safer place. Exactly the opposite is true. We have to figure out what is most essential in the protection of these kinds of resources, and stop pretending that we can protect them all. About the only two things that will offer such a broad range of protection are good intelligence and deterrence.

The resources devoted to collection and analysis of our adversary's capabilities and intentions do not match our concerns for cyberwar. Anytime attribution is the biggest impediment to identifying where a hacker is coming from, we don't know enough about who is doing what in the cyber world. Are those Chinese hackers coming from China, or are they Russian hackers coming through Chinese servers? We should know. We spend too much money on intelligence every year to not know.

Deterrence is a different matter. How do countries provide deterrence to cyberwar without performing the same kinds of infrastructure penetrations that we are experiencing? That worm in the electrical control systems or those banking systems would not be of much use if nobody knew about it, so the game is very difficult to play. We have to be able to show our opponents what we will do if they continue, but we will have the same problem with cyber that we have with nuclear weapons. If North Korea launches the attack, do we assume China is behind it? Do we retaliate against China? It isn't easy to figure this out, and the issue isn't settled yet. It still goes back to good intelligence, so we can know whom we need to deter.

Our main deterrent capability is the chance that Western countries are prepared to give back what they get in this type of battle. That has limits. Someone like ISIL, given the opportunity, would certainly do as much damage as possible and disregard the collateral damage that we might do to them. It would all be good (for them). That means being prepared to take down the electrical systems of some pretty big countries or attack their capability to assault other countries by exposing their plans or preparations. We seem to be reluctant to do either.

This is total war. We won't see it coming, and they probably won't either, but we shouldn't have to guess. We should know.

If Stuxnet was any indication, our enemies should worry about the possibilities. The West has capabilities it has never shown, and would not want to show an enemy until the time is right. Only that kind of deterrence will keep them from what will amount to a very interesting war.

We do not have deterrence sufficient to deter China and Russia from stealing information from us on meddling in our internal affairs. That means we don't have a deterrent strategy at the national level that works. We have to make both countries believe that the consequences of their actions will have an affect they will not like. So far, the U.S. has not demonstrated that capability.

If we want to attack a network (technically known as Computer Network Attack, or CNA), it will be with the intent to do damage to its function. So if Mongolia decides it wants to cause trouble in Moldavia, it can attack its telephone system and stop Moldavians from making calls on any landline or cell network. In this scenario, the Mongolians can choose to use their military, a contractor, or a government agency not related to the military. In some countries just that choice sometimes eliminates the military from participation.

But if Mongolia wants to collect intelligence from the Moldavians, it would need to exploit the telephone system (Computer Network Exploitation, or CNE). It might get into the cell network and steal every conversation between certain people in the government. That raw data would be analyzed and stored as intelligence for use by the Mongolian government. This intelligence gathering is usually done by an intelligence service, or one of its contractors, although that doesn't exclude the military altogether. Military units support these kinds of operations in many countries. They will cooperate on these operations given different skill sets that might be available to them. Even where no military is involved, they share data with the military units regarding things that might affect them.

Then, someday, Mongolia will want to make strategic decisions on how to use this information. The Mongolians might play some of those calls on television to embarrass an official who is about to run for office, or they might even blackmail the caller into a change in behavior more suited to their cause. Neither of these options generally involves military participation.

Mongolia might have to devote several years to getting into the phone system in Moldavia before it can completely shut the system down. We have a misguided view of how simple this supposedly is because television tells us how easily we can hack into almost anything and produce a devastating result. The really good TV hackers can do this in a couple of minutes. If it were that simple, we would not have an Internet, because governments would not allow it.

Suppose Moldavia and Mongolia both use the same carriers. We might have a company like Telefónica or Vodaphone operating there. Mongolia would have to consider that neither of these large companies wants someone conducting operations against another country using their network. They might like it even less if they are attacking another one of their customers. Mongolia would have to disguise its operations to look like something legitimate, and this takes time and money to do effectively.

Mongolia might have to use the same techniques to get the cell-phone conversations from government officials. If it plays those for the press soon after it gets into the network, the Moldavians will stop using the cell phones and improve their security, just as the Russians did after Chechnya. Politicians, who may know about the effort to collect information, are often prone to announce something while they are in office to help them stay in office, and believe it is urgent to release the information. Senior government officials may not think it is as important as the intelligence they are getting about Moldavia's next election. They have to control their politicians better than most countries are able to do.

The Mongolians can't be too quick to tell everyone they have this data, or they will be seen as the source of the story. The difference in intent lies in how the attacks are being used to collect information or damage the capabilities of another country. Neither of these is a simple thing to do, and they cannot be done quickly. The military units may not be involved unless the attack is a prelude to something else.

What the Chinese did at the *New York Times*, and what *Forbes* says Syria did to it (for the same reason), was Computer Network Exploitation. The hackers weren't trying to stop the *New York Times* or *Forbes* from operating a news service, so they didn't attack their systems to disrupt them. They could have; they just decided not to.

A network attack might have worked for a short time, but in the long run, news services will still be around when they figure out how to stop it. The Chinese were trying to exploit the *Times* system to find out who was giving the reporters information for some of their stories. The longer they remained undetected, the more success they would have in identifying the sources. Exploitation becomes harder after the host finds out it is being exploited and starts to do things to prevent or detect incursions.

11. A Military Left Out

The indictment of a Chinese army unit mentions five companies and one trade union, but it does little to define the scope of collection in the United States. The Justice Department cut a deal to limit companies named in the indictment to only those that agreed to be named. This People's Liberation Army unit may be the group responsible for a hefty number of attacks on U.S. industries (like Coca Cola), government agencies, and individuals. They also focused on gas lines, the electrical power grid, and waterworks, "including one company [later identified by Mandiant as Telvent, a company that maintains blueprints] that had access to 60 percent of oil and gas pipelines in North America."[3] But they are far from the only group operating in China. There are more there and in other countries, and their work is about to get more interesting.

Three years ago, Greg Kipper at Verizon told me that the Internet of Things was coming. Hearing about it and visualizing it are two different things. It was a concept that was hard to get a mental image of, since it is not just the Internet in the conceptual framework we typically put around it.

Kevin Ashton of MIT and Auto-ID claims to have originated the term in a speech at Proctor and Gamble in 1999, and says that even though he used the term first, he has no right to define how others might use the term today. But what he meant by it was as follows:

> If we had computers that knew everything there was to know about things—using data they gathered without any help from us—we would be able to track and count everything, and greatly reduce waste, loss and cost. We would know when things needed replacing, repairing or recalling, and whether they were fresh or past their best.
> We need to empower computers with their own means of gathering information, so they can see, hear and smell the world for themselves, in all its random glory. RFID and sensor technology enable computers to observe, identify and understand the world—without the limitations of human-entered data.[4]

The Internet used to be connected to people on computers that were fixed to a desktop. Ashton rightly points out that most of the information on the Internet originally came from human beings. It was processed, managed, and distributed by computers, but it was still human-originated information. He wants to broaden the use of computers to collect information in their own right. If this conjures up thoughts of Skynet and impending doom, so be it, but computers are doing this kind of thing on networks every day without consequence. Computers now manage the Internet with very little human intervention. Computers monitor Internet traffic, and when the volume exceeds norms, adjustments to data routes and priority of messages are made without reference to humans. Computers are learning to drive cars without human intervention and without wires in roadways or a dependence on every vehicle on the road being computer controlled. They are learning to "observe" the environment in which they operate and to react to changes in that environment faster than humans. Computers can be given parameters in which to operate that will not be ignored; you can't do the same for people.

The Internet is a worldwide telecommunication system that provides connectivity for millions of other, smaller networks; therefore, the Internet is often referred to as a network of networks. It allows computer users to communicate with each other across distance and computer platforms.[5] The United States has the most say in how the Internet is operated, much to the chagrin of other countries that wish it were not so, but the current administration recently discussed giving up this control. The technical core of the Internet is managed by a private-sector, international organization called the Internet Corporation for Assigned Names and Numbers (ICANN). It may be difficult to figure out what ICANN does just by looking at its name, but it manages and oversees the two most important technical areas—Domain Name Systems and Internet Protocol Addressing.

Before the Internet grew up, it had the names and network addresses of each computer in a single file. If I had typed amazon.com into my computer, it would have looked for the Internet address of Amazon and taken me there (had there been an Amazon then). There are far too many computers on the Internet these days, and some 30 *trillion* websites, so the administrators developed a distributed system to handle the name and address conversion, which works both ways. It was not designed to be a secure system, since the developers assumed the users were all of the same mind. That type of thinking works well until a number of places start masquerading as Amazon.

When we search for Amazon now, we get a whole list of options that are Amazon, relate to Amazon, or are pretending to be Amazon. A few of us might pick one of the pretenders in searching for the real one. Most people have an Internet security package on their computers that looks for sites pretending to be Amazon, as well as ones that identify sites as advertisers on Amazon rather than the real Amazon. Since the Internet Domain Name System doesn't validate the fact that Amazon has a set of addresses that are for its exclusive use, it is possible to get one that is set up to trap my credit card number. It might even log into Amazon as me, make the order I intended, and then go about using my card number until one of the banks catches up with it.

This has become a familiar issue because of Heartbleed, a vulnerability in the construction of the software that establishes a secure link between a user and a website he connects to. Hackers are making it more difficult for our defensive systems to operate by stealing valid encryption certifications to websites and going after other forms of user identification. When we communicate with our intended websites we think they are secure, and they aren't. Vendors are having us change our passwords to make sure we can get into

11. A Military Left Out

their sites, and we have come to know the problem through that requirement. When you change your password on a legitimate site, the stolen credentials stop working. The Internet is not as safe as commercial businesses would like us to believe, and it is becoming less so.

The militaries of the free world have a two-fold problem. First, they don't own the Internet, nor are they responsible for its security. Second, they can't stay off of the Internet even though it was never intended for their use.

The Internet does not have a genuine police force to supervise it. We note that China has one with little cartoonish characters reminding citizens of its presence. Russia also has one, which is supposed to be looking for child pornography in a country that sells it more than any other. The free world has police who look at criminal activity that might take place on the Internet, but the difference is that they look for the same activity whether inside a computer or outside where we can see it. They are not, strictly speaking, Internet police, but they are police who specialize in digital crime or who use the Internet to identify and catch criminals.

The basic problem for the militaries of the world is an inability, or unwillingness, to separate from the Internet. Because the U.S. Defense Department had so much to do with the establishment of the basic architecture of the Internet, computer security policy came from the military in the 1970s. Gradually it became less a military network and more of what we think of when we say Internet today.

The U.S. military has something called NIPRNET (Non-classified Internet Protocol Router Network) that costs millions of dollars a year to maintain. This unclassified network is supposed to handle all the unclassified work of the military so it doesn't have to use the Internet. However, it isn't considered safe enough to perform these government functions.

When Chinese hackers got into NIPRNET in 2007, the Defense Department downplayed it as a network carrying only unclassified information, but if this type of data were not valuable, there would be no reason to have a special network to put it on.[6] A year later, they had a real reason to be concerned. Agent.btz (the same one used by Russia in the Ukraine) showed up on the classified networks of the department.[7] At the same time, the United Kingdom's Ministry of Defense (MoD) was saying it was concerned about attacks against its top-secret networks.[8] The minister added that these systems were not connected to the Internet.

Year after year, the Government Accounting Office (GAO) and the respective Inspectors General of the military services cite poor performance of security practices in their networks, but year after year very little is done about it. These are the most sensitive, important networks anywhere in the world, but the operators haven't done enough to make them secure. (We can include their contractors too.) Now the GAO has even more to fear.

Everything is connected to the Internet, including those military systems that are (supposedly) never connected to it. These aren't lies being told about their connectivity; our leaders actually think they aren't being connected when in fact they are. In the same way, through commercial penetration testing of sensitive business networks, leaders were always surprised by connections made to some of their most restricted data. Before our audits, their technology staff would say their networks were isolated and very secure.

If networks were static, we could say with certainty that something was not connected to the Internet and be sure that when we went back to check, a year later, that it would not be. In our networks in Ballistic Missile Defense we had very active network mapping tools that, every few minutes, we used to get an updated status on network components. We were occasionally surprised, in this very constrained environment, at what popped up on that status board. In business, commercial network mapping was even more surprising.

A few years ago one of our clients was not happy to find that a part of his network intended for R&D was connected to the Internet. R&D was an area that was supposed to be isolated, and even some of the internal technical staff did not have access to it. The information on over 20 patented lines of work was being tested there. When we looked at why the network had become exposed, to literally anyone who wanted it, we found a test server that was used early on in an experiment to link the network to another company. The other company had decided it was too risky to connect until it had some robust encryption established on the link. The test team subsequently forgot about it, and it was still there a year later, with no encryption or a connection to the other company. It was just a wide-open path in that was quickly closed.

For owners of top-secret networks, like the ones in MoD, it is hard to imagine that there is any path for any kind of virus or worm to get in from the Internet. But that happens a little at a time, and usually "by accident." The top-secret network connects to a secret network, and there are interfaces between them that are supposed to keep the networks secure from each other. The secret network then connects to a network like NIPRNET, which connects to the Internet. In the early days of U.S. computing, that layer of connection was not allowed (i.e., there was a policy to prohibit any connection, even an indirect one, to the Internet). MoD, DoD, and hundreds of other agencies can't, and won't, stop connecting to the Internet even though their governments pay for considerable capability intended to prevent just that.

What makes the situation worse is the connection of contractors and business associates in both government and private businesses, who each have their own connections to various things on their networks. That sometimes includes Internet connections. The Chinese and Russians don't have to attack their targets directly, and they don't. They don't hack the *New York*

Times or DoD; instead, they can hack a business partner and use that partner to get access to the target. Hackers generally gravitate to the least well-defended network, and every business will say it isn't them.

The interconnection of multiple networks is what makes any statement about what is connected to the Internet a guess. It grows too fast to keep up with. Any connection is the next way in, and hackers will take advantage of it in ways the technical staff never imagined.

For every escalation in defense, there is a new attack developed. If we look at a current attack, originating in Syria, described by Andy Greenberg of *Forbes*, we can see how hacking by nations differs from hacking by groups of people who are trying to make money. According to the *Forbes* account, "Hacked by the Syrian Electronic Army," March 2014, the attackers went after a connected site at Vice Media. They compromised the site, and then sent an e-mail to some of *Forbes*' employees and told them a story was posted on Vice Media that would be of interest to them. When *Forbes* staff members logged into Vice Media, the Syrians got the credentials and used them to get into the *Forbes* network.

This situation is about to get more complicated because connected to those networks are such odd things as self-guided cars, smart buildings, power station control systems, dams, banks, ATMs, ordering systems, bridges, sensor displays to see and hear what is going on around a home or in traffic, tax records, medical histories, browsing histories, private memos, communications with other people, news services (and accompanying images), refrigerators, TVs, telephones, ovens, security systems, snack machines, collections of buying habits, classrooms, and independent education opportunities at MIT, Harvard, and elsewhere. There are so many things connected to networks that it is hard to visualize all of them. They contain a mix of human and computer-collected data. All of these have potential paths to a number of other networks and provide an enemy with plenty of ways to attack. The military has nothing to do with defense of any of these paths in.

This is a set of infrastructures that are privately held, running on private networks, and increasingly vulnerable to attacks by governments. Think about Google, Microsoft, Facebook, Verizon, AT&T, Sprint, General Motors, Hewlett-Packard, Amazon, and hundreds of other entities, with their own networks so large they are hard to imagine, asking for help from the military to defend these networks. Not only are they not going to ask for that kind of help, but they are probably going to reject the idea before that discussion goes very far inside the organization. Since the military doesn't own their networks, it really has no business trying to defend them.

From a user perspective, the kinds of attacks that governments are conducting are impossible to defend against. Large companies find it difficult with far more resources. It may take several years to accept that, as one person, there is very little that can be done to stop hackers from getting your work, reading e-mail, watching social media activity, and monitoring where you go on the Internet. *Get over it* would be the best advice. It isn't going to improve. The best thing we can do is recognize that our private conversations with people on the Internet are never going to be private.

Even if we do all the things that we know can be done, there is still very little chance of success unless the carriers, application builders, cloud maintainers, network providers, shared resource providers, and equipment manufacturers all do their jobs to the same standard, with the same diligence. Google and Apple now will use encryption to try to help us. We have to trust them to implement that system so it protects us. We can't do their job for them, and most often, service providers are the weakest link.

Our militaries have always been the sector of government we go to defend us from foreign threats, but this time cyberwar (at least in the free world) is being waged by personnel other than military units. The non-military units are hopelessly divided and incapable of dealing with government-managed attacks. Edward Snowden may not be a popular spy, but he did one good thing by allowing the world's literate population to discover what happens when we turn loose a dedicated group of technically competent people on the infrastructures of electronic devices. Not even the best of defenders can stop them.

12

The Integration of Business and War

The business of war is first and foremost the use of business objectives to impose the political will of one government on another. In a lesser context, it is the integration of business objectives into the objectives of war. Most, but not all, of this activity is part of economic warfare.

The Chinese are the leaders in modern war. What separates them from the rest of the world is that they have a government that combines business, the military, and the state. There is less separation of goals and strategic objectives, and the government enforces its policies through a series of state offices located in private businesses. The Chinese carefully control their Internet and internal communications, and they steal from those who don't. They have nationalized industries that operate under government control, many by the military. They use their national intelligence assets to help their businesses grow, and their power at the top is controlled by the Communist Party. At that level, the party is small and speaks with a single voice. Contrast that with national governments in Western countries, and it is not hard to see why China does well when competing with the rest of the world.

The Russians are different and slightly behind the Chinese, but they have business oligarchs surrounding the imperial leadership. That brings the business objectives into alignment with the political objectives of the government. This aspect allows the Russians to do things like manipulate gas prices in the Ukraine, raising and lowering them as political winds shift in Europe. They doubled natural gas prices "because the Ukraine was unable to pay its bills" and was a greater risk. The Europeans are helping to settle the issue because they are on the end of the pipeline that keeps the Ukraine gas flowing and fear a cutoff.[1] The Europeans have seen all this before, and they didn't like it very much the last time.

In 2010 WikiLeaks published a number of cables from U.S. diplomats describing Russia as a "corrupt, autocratic kleptocracy"[2] (a kleptocracy, according to Dictionary.com, is a government or state that exploits natural resources to steal—rule by a thief or thieves). This is not diplomatic language, and while they might have been writing in classified cables, they should have known better. Defense Secretary Gates characterized it this way: "Soon after the Serdyukov visit, I had told my French counterpart, Alain Juppé, that democracy did not exist under Putin, that the government was little more than an oligarchy under the control of the Russian security services, and that although Medvedev was president, Putin still called the shots."[3]

While the diplomatic cables were even harsher than the Defense Secretary, they were really talking about two different things.

Russian corruption is a concern for any business working there, with some questioning whether business can be conducted within suitable standards of conduct. Western businesses "routinely pay bribes to Russian government officials" in spite of several anti-corruption initiatives.[4] Most people convicted in Russian courts for bribery offenses are government officials. Transparency International says the perception is that Russia's corruption has increased in the last three years. A small Finnish window company took everyone to court who came to its facility asking for money; when it got to 44 convictions, the visits stopped.[5] This example serves to illustrate that there may be some relief in Russian courts.

What Secretary Gates said about participation by the FSB has nothing to do with bribery. He was instead talking about the Russian government being run by former members of the security services with close ties to the current FSB at different levels. This would be comparable to having most central U.S. government offices populated by former CIA officials. BBC cites a study by the Russian Academy of Sciences that says participation of the security services is increasing in government since Vladimir Putin took over: "Among the presidential administration, members of the government, deputies of both chambers of parliament, regional heads, as well as the boards of Russia's top state corporations, four in five officials worked for the KGB, or continue to work for one or more of its successor organizations."[6] The emphasis on intelligence agencies gives a higher priority to broad intelligence collection, analysis, and covert operations.

Increasingly, governments appeal directly to an opponent's population, bypassing the government whenever necessary. Where possible, they do it covertly. What the Russians did in the Snowden espionage incident was use information he collected to directly appeal to citizens of other countries through their press outlets. The press did the translations for both language and culture. The Russians influence existing leaders and thought leaders to accept views consistent with their own. Failing that, they criticize. They only use force, like they are doing in the Ukraine, in areas where agreement is not possible.

12. The Integration of Business and War 143

In quite a few ways Russian crime fits their business model. Criminals bring in cash, which becomes part of the economy. In one recent incident in Australia, a Russian gang made off with nearly $570M.[7] Criminals go for big rewards outside their borders and, unlike the Chinese, avoid attacking their own country so they don't become a threat to the Russian government. They are not political and don't make public displays of political agendas.

The perception of the mingling of Russian business and criminal figures may have come from a Spanish judge, an expert on the Russian mob, describing the relationship between the Russian Mafia and central government as "a mafia state."[8] So, while we might be concerned about the creeping FSB participation in the Russian government, we might be even more concerned about their takeover of these criminal rackets. A potential refugee in New Zealand demonstrates how this might be happening.

In 2011, a Russian applied for refugee status, citing less than satisfactory relationships with two individuals—an FSB officer and a senior member of a Russian Organized Crime Group (OCG). He and two others had started a business to import and sell alcohol. In order to establish and remain in business, he had to pay *krysha* (roof), or a kind of bribe to regulatory agencies. The amounts of the bribes were based on the size of the business, and by the time he left Russia he was paying $10,000 a month. If he had not paid, regulators would have said his business had rats in the warehouses (or some other made-up violation).

Seven years before he decided to leave his country, he expanded the business and hired a known OCG member, with connections to a big OCG in Moscow. He thought this would immunize him from dealing with several smaller gangs. Instead, he was approached by the head of the local militia, who demanded $3,000 a month for his services. (Note: The Russian national police were known as militia until 2011.) When the man refused to pay, he found out what those services were; they closed one of his warehouses.

He appealed to the local magistrate, but before his appeal could be entertained an investigation was required. It dragged on for several months, while demands for more money continued. When he finally got his keys to the warehouse, $220,000 of his alcohol was missing.

[Editorial note: Earlier it was noted that the Finnish window maker was able to take bribery suspects to court, where they were convicted. In the situation of the alcohol importer, however, matters became worse, not better, when involving the courts. This may be explained by the way the Russian legal system works. Russia has very few lawyers, so citizens often represent themselves in what we would consider civil matters, like disputes between two individuals. Russia does not use case law as in the United States. Each judge or magistrate reviews the law and makes a decision based upon his or her interpretation of the law. There is no appeal process per se, but if an upper court receives too many complaints about how a law is interpreted, it is able to issue a finding indicating how the law should be interpreted. The upper court is not obligated to issue an interpretation, although it has the authority to do so.]

Later that year the man was visited by another leader in the local OCG, who demanded a payment of $300,000; he refused. He called the local militia, whom he knew well by that time. They arrested the second OCG leader and some members of his gang and held them for a few hours, but didn't charge them.

Three weeks later masked men with baseball bats attacked him in his home. During the fight he got the mask off one of the men and recognized him as one of the second OCG's gang members who had been to his house before. He reported the assault to the militia, after which he was advised by the leader of the first OCG that he had made a mistake in reporting the incident outside the local OCGs. They had to resolve the matter internally, and this created complications for them.

Charges were filed against the man for selling alcohol after it was expired and he was jailed. While in jail he was visited by a senior FSB officer who said the Russian was to sign over his business to him. He did not. He was later threatened and one of his guard dogs was shot at the warehouse. Shortly afterward, four armed FSB officers attended a meeting with him to attempt to find out what was going on with their fellow officer. There was discussion, but the meeting broke up without any arrests.

The prosecutor's office and tax authorities started demanding bribes and his son died "in a suicide," so he decided a move to Moscow might be helpful. It wasn't. The OCG there wanted a different business model, so he went back home. Shortly after he returned he started to receive death threats from an FSB officer, inspiring him to apply for a visa to New Zealand. It was originally denied, but he won on appeal.[9] The Russians would argue that a person applying for refugee status would say anything to obtain a visa, but the New Zealand government believed him. His description was long, detailed, and consistent with other cases they had seen.

[Editorial note: In early 2000 the Russian Federation for Legal Reform (RFLR) contracted with my employer, KPMG Consulting, to determine the most economical way of communicating current law to Russian citizens. This study was conducted after the decentralization of power and the subsequent withdrawal of some territories/countries from the Russian Federation. Few Russian citizens outside major cities had access to communications other than newspapers; Internet access was extremely limited. However, during the Soviet years the Communist Party had headquarters in nearly every city, and each headquarters building was equipped with communications equipment that was "modern" by Russian standards. As noted by the author, this was a way in which the party controlled the message—all services and communications were controlled by or through the local government headquarters. What remained after 2000 was a government facility with communications capability, including, in most cases, access to the Internet. Because citizens were conditioned to visit party headquarters—"the government"—for services and communications, it seemed natural for the "new" Russian Federation to take advantage of these distributed information hubs to provide access to current law for Russian citizens; as far as I know, they adopted this model.

The transition to the Russian Federation and reduced dominance of the Communist Party did little to change the power structure outside of Moscow. The change increased autonomy of local government and gang leaders; Moscow was busy defining and organizing the Russian Federation and had little time for concerns in individual communities. Life for the average citizen did not change significantly. The FSB remained in each community, as did the OCGs; the same local government officials were bribed when something was needed. Controlling communications was key to local government officials retaining power during the transition from the Union of Soviet Socialist Republics (USSR, or CCCP in Russian) to the Russian Federation. Until Putin began reconstituting stronger centralized control, local governments and courts were autonomous as long as they did not attract attention to themselves. Local autonomy allowed corruption to flourish unchecked; in some cases, it allowed a more entrepreneurial way of doing business to flourish. The alcohol importer made the mistake of establishing his business in a highly corrupt area where the magistrate was apparently as corrupt as the local government officials; the Finnish window maker did not make the same mistake, or else was just lucky. The attempt to elicit protection from a large, non-local OCG might have been logical in different circumstances, but it was not logical in post–Gorbachev, pre–Putin Russia.]

China and Russia both have the same kind of problem with corruption in their militaries stemming from a mixture of business and government. The Chinese agencies responsible for oversight of businesses are often the ones responsible for their business direction. For example, Ministry of Industry and Information Technology runs the carriers, China Mobile, China Telecom, and China Unicom, issues licenses to them and controls investment categories through the Industry Catalog and the Provisions on Administration of Foreign-Invested Telecommunications Enterprises. They regulate and manage investment at the same time. In Russian business the oligarchs and government offices supervising them are influenced by the need to reduce competition and control ownership, resulting in the same end. China has started a long process of eliminating some of its corruption by making examples of powerful people. President Xi Jinping removed General Xu Caihou, one of the most senior officers in the People's Liberation Army, in 2014. He was on the Chinese Politburo and a vice chairman of the Military Commission.[10] Another general, Gu Junshan, the former head of PLA logistics, was arrested at the same time. General Junshan's house in Henan Province was shown on television stations as a symbol of corruption and an example of a person living beyond his means; he couldn't have been that rich on his salary alone. Charges have yet to be filed, but the military in China is a very large and powerful one and intermingled with its economic war. It operates a number of businesses, where much of the collusion starts. Military leaders are allowed to own and operate businesses and keep some of the profits. Taking that away may lead to more than just a mild change in behavior, so President Xi is walking a fine line.

Xi took an even bigger step in 2014 by detaining Zhou Yongkang, a former member of the Politburo Standing Committee, the highest decision-making body in China.[11] Zhou was the head of the organization that includes China's police, prosecutors, judges, and intelligence services. He has yet to be charged, but the investigation is a year old and they usually don't rush this kind of action.

Russia's crime problem in the military also stems from back-door corruption that is officially recognized and rapidly increasing. The current Defense Minister points to his predecessor as part of the problem. He was fired in November 2013 and questioned by prosecutors after he "lost" military property valued at 1.3 billion rubles. Several investigations followed and multiple-million-dollar cases are being reviewed.[12]

All of the cases involving Russia's former Defense Minister involved improper dealings with defense contractors. In 2011 Dmitry Medvedev, then the Russian President, fired several defense officials for "failing to fulfill state contracts," reminding them that in Stalin's time they would have been worse off. Nikolay Petrov, an analyst at the Moscow Carnegie Centre, said the crackdown was intended to show that the Kremlin was in charge and taking care of the armed forces.[13] The head of a nongovernmental anti-corruption committee said 40 percent of the military contracts are not delivered or are of lower quality than expected.

The corruption and splits between leadership and the armed forces could leave a finger on the trigger that might have an inclination to work on improving its own position, and not the position of the government as a whole. This might be short of the *Dr. Strangelove* image of doomsday, but it could still trigger a good bit more anxiety between countries than any of us would want.

What the Chinese have done better than most others is integrate their businesses into their governmental strategic goals. Their businesses follow the government's lead or the business gets new leadership. Their focus is not on profitability and shareholder returns (something stockholders in Chinese companies should realize). Instead, they use businesses to achieve war objectives, something the rest of the world is beginning to consider.

Russia has done something similar, albeit using different methods. Both come back to central management of the government and business by small groups of authoritarian leaders proclaiming democratic principles. They are really talking about power—staying in and keeping others out. Control over

12. The Integration of Business and War

the military, government, and business gives both countries an advantage in dealing with democracies. Their businesses and governments are focused on similar objectives, consistent with maintaining power and winning the will of the people. If they were content to control only their own people, we might be more tolerant of their behavior.

If we use modern war to achieve our objectives, von Clausewitz was not far off in his 1800s definition. War has become a dirty word, like bullying. Nobody wants to be at war and, as we are finding out, nobody has to be.

The U.S. military thinks of information war more as an adjunct to real war, something that will augment a country's ability to deal with aspects of a war that can't be dealt with any other way. Our enemies look at it as a way to win a war without ever being at war. Both sides right about their approaches. Where we have difficulties is when businesses make demands for war materials without help or guidance from a government.

When President Dwight D. Eisenhower was about to leave office, he gave a speech to the American public. It was about something he called the military-industrial complex. In his farewell address, he told his 1961 audience that a storm was coming and we needed to be concerned about the complex of our military and contractors that came out of World War II:

> Until the latest of our world conflicts, the United States had no armaments industry. American makers of plowshares could, with time and as required, make swords as well. But now we can no longer risk emergency improvisation of national defense; we have been compelled to create a permanent armaments industry of vast proportions.... We annually spend on military security more than the net income of all United States corporations.
>
> This conjunction of an immense military establishment and a large arms industry is new in the American experience. The total influence—economic, political, and even spiritual—is felt in every city, every State house, every office of the Federal government. We recognize the imperative need for this development. Yet we must not fail to comprehend its grave implications. Our toil, resources and livelihood are all involved; so is the very structure of our society.

At the end of World War II, 16 million had served in armed forces of the United States, and by 1944 there were about 8 million men and women in the army alone.[14] The United States is now debating whether it should cut that number to less than 500,000. We don't really seem to know what the right number should be. The personnel strength of any military should not be calculated based on economics. We should know how many people are needed and fund that number.

We have fought a series of wars since World War II, but we only have a fraction of the forces we had then: 2.2 million uniformed military and civilian employees in all of the Department of Defense.[15] Yet in FY 2012 the DoD obligated $360 billion for contracts for acquisitions—more than all the other government agencies combined. Jacques Gansler, who was the Deputy Secretary of Defense for Acquisitions when I worked in the Pentagon, said he was seeing the department issue 52,000 contracts every day. Over half of the people supporting the military in Afghanistan are defense contractors.[16] Contractors perform essential services—far too many of them.

This year Congress authorized money for Abrams tanks that the army says it has enough of. The navy is keeping seven ships, even though it wants to mothball them.[17] Congress won't let the air force retire the C-5A Galaxy transport plane, while the air force says there is no money to fly these planes, put pilots in them, or even repair them. The planes sit on the runway in Texas, moving now and again to keep the tires from rotting. Congress also won't let the military retire the Global Hawk surveillance drone or C-130 cargo planes.[18] The sad thing is, we could have operated a large network of security specialists needed to protect NIPRNET from the Chinese with the money that just one of those tanks cost. Shifting priorities to cyber means not building so many weapons we don't need.

Tanks, ships, and aircraft the military does not want, still in our inventory of weapons because of something that has nothing to do with military capability. This reminds me of the airborne laser, which was a missile defense system mounted on a Boeing 747. In a meeting I attended in the Pentagon, one of our weapons experts was asked how the test bed was coming along. She said, "The only way that thing will hit a target is if the pilot flies into it." That was 10 years before the decision was finally made to abandon the project and mothball the airplane.[19] In 2010 it proved her assessment wrong by actually hitting a target—with the laser, not by flying into it.

From time to time Congress votes to keep military items in the inventory even though the services have asked for them to be discontinued or cut back. It is no secret why we continue to have congressmen voting for weapons nobody wants; it's politics. Unwanted weapon systems go back to the ties Eisenhower feared between the military and political leaders of the country. If we ask who benefits from keeping weapons in stock that the military does not want, it certainly isn't the people who rely on the military for their safety and security.

It matters because the defense budget is a zero sum game. If we spend money on weapons the military doesn't want, it can't shift money into cyberwar and has to reduce its personnel that fight the new wars. This is something of a dichotomy: We have steadily reduced our armed forces, yet the cost of supplying them with weapons is not going down nearly as fast as the manpower. We build weapons the armed forces do not even want; we build ones that we already have too many of, and we don't have enough money to pay some of our soldiers and airmen. They sometimes buy their own body armor to go to war. It is a dichotomy that is hard to explain to parents whose sons and daughters are serving in our armed forces.

12. The Integration of Business and War

In the United States, the military is not supposed to influence Congress directly regarding appropriations for weapons it favors. Some Secretaries of Defense, like Donald Rumsfeld and Caspar Weinberger, were adamant enforcers of this policy, and others weren't. I served in the Pentagon, and later on the Hill, seeing both sides of that influence. Though they could get themselves in big trouble, the military leaders were constantly trying to get money for their own systems (most of the time with help from a congressman or two).

Almost every day a few generals make their way up to the Hill for private meetings with staff, congressmen, and sometimes each other. They are all looking for support for weapons of one type or another, typically made by a company in the state of the congressman they are visiting.

The generals are almost always "called to discuss," like they would never go on their own, but someone asked them to come and they can't say no. Congressmen, staff, the military, and contractors expertly play this game. They draft amendments favoring one type of weapon over another, purchases of support equipment, or services performed by specific people. They kill certain types of legislation that might favor something else that competes with them. They give presentations, support documentation, test results, testimonials, and lots of parties that stretch to all hours. There were staffers who went to as many of these parties as they could because they couldn't afford the same quality of food on their own salaries.

Every administration, and every Secretary of Defense, has made an effort to enact acquisition reform to slow this growth.[20] They are not likely to succeed because reforms won't lower the cost of defending our country. The defense contractors are part of the business of war, and that business is greater than just war machines. They sell skilled human resources, political contacts, research, and a host of "services" (which can mean almost anything).

If we seem to have more arms than we need, it may be because we have more arms sellers than we need. It is a global business. Some of the biggest contractors are not U.S. companies, and some of the biggest U.S. companies are selling overseas in competition with those already there. BAE Systems Inc.; Airbus, which is the name for the European Aeronautic Defence and Space Company N.V. (EADS); Finmeccanica; Thales; Safran; Rolls-Royce; Almaz-Antey; Mitsubishi Heavy Industries; Saab; and Rheinmetall—while not exactly household names to a lot of people, these are still among the top 25 weapons systems makers in the world.[21]

BAE and Airbus are fairly well known for making airplanes and armored vehicles, along with many other things. Both actually make aircraft. Rolls-Royce builds aircraft engines for almost all the major airlines, both military and civilian. Finmeccanica, partly owned by the Italian government, specializes in helicopters. Thales, a UK company (and one of two finalists in the competition for Poland's $5 billion missile defense system), builds combat management systems, as well as naval, airborne and ground intelligence, surveillance and reconnaissance systems, and it has electronic warfare capabilities. Safran S.A. is a French company specializing in propulsion systems for various vehicles and aircraft, defense items, and airport security equipment. Almaz-Antey is a Russian joint-stock company that does research and development and makes defense missile and radar systems for land and naval use. It also makes TV transmitters, telecommunications systems, air traffic control and radar. Mitsubishi Heavy Industries (MHI), similar to BAE and Airbus, is a large conglomerate of companies (in MHI's case, 294 all over the world). They do space launches; build airplanes, trains and ships; and provide services connected to almost everything in defense or transportation. Saab makes aircraft and provides services related to defense and force protection, electronic defense systems, early warning, civil defense, training and simulation, telecom carrier, and power solutions. Rheinmetall is German with specialties in automotive technologies and defense systems that include mobility, reconnaissance, command and control, firepower and force protection. It builds armored vehicles, weapons and ammunition, air defense and electronics.

The biggest are all U.S. companies: Lockheed Martin, Boeing, General Dynamics, Raytheon and Northrop Grumman make up about a third of all sales. Worldwide the arms manufacturers sell about $710B a year.[22]

Almost none of these companies work alone. They have a series of interrelationships that allow them to build big things in large quantities, which smaller companies cannot do. They feed from the same sources and work together. About 10 percent of the companies that do defense work have 90 percent of the business.

Lockheed, in 2012, got approximately 17 percent of its total consolidated net sales from overseas.[23] Half of the amount was through the U.S. government as foreign military sales (FMS); the other half was direct sales. Northrop does 90 percent of its business with the federal government, so that doesn't leave much else. Raytheon gets almost 26 percent of its sales from overseas, partly in FMS.[24] However, FMS costs don't reflect the total cost to the government for these contracts. The Defense Department pays for part of its workers' pension costs for work expended on military programs. Billions of dollars will be spent on these reimbursements in the next few years.[25]

Foreign military sales are about $69B a year.[26] In 2012 the bulk of those sales came from the Saudis for Boeing's F-15 fighters. Saudi Arabia spends proportionally more on defense than any other country, with Israel second.[27] However, that doesn't represent what goods and services are provided to other countries' militaries. The total for that is in the $394B range, according to the U.S. Defense Security Cooperation Agency, which coordinates global security cooperation programs, funding, and efforts across the Defense Sec-

12. The Integration of Business and War   151

retary's office, the Joint Staff, the State Department, the combatant commands, the military services and U.S. industry.

Sometimes, these defense contractors are big enough to make their mark on foreign policy. They lobby Congress and legislators of other countries, trying to influence appropriations for weapon systems, exports of goods to other countries, or the opening of international facilities to move technology where they think it will make them the most money. All those teaming arrangements help to make them big—and influential.

In a study of the potential areas of influence on foreign policy, two researchers found business leaders had the most impact, even more than labor leaders; the general public has "little or no effect." We may believe or hope that popular opinion influences our leaders, but in reality they don't pay much attention to us. This study also found that business influenced who was recognized as an expert and what those experts said about the issues.[28] The industry defines who the experts are that speak for it, and it frames the messages to fit what they want to achieve. Those talking heads we all see on TV are not just any heads brought in off the street.

We might be skeptical about this claim because our army is shrinking to numbers that make us nervous, but among defense contractors there is plenty of excess capacity. There is also more to a standing army than its personnel, and this is more to the point than human strength figures. We have too many weapons for too few people, and the ones they have are not nearly as good as the makers profess. The combination of the use of power and low personnel numbers required more tours in Iraq and Afghanistan for our military personnel; more tours means more exposure to threats. When someone dies in war after their third or fourth tour, we shouldn't be surprised. Exposure amplifies the risk.

Crony capitalism is a term most associated with banks and insurance companies that are "too big to fail."[29] Governments support these financial companies because they are considered important to our national interest. The total should be small, as the banking crisis showed, but the cost of reducing the number of too-big-to-fail banks and insurance companies is significant. What the Federal Reserve found was that banks that were propped up and "too big to fail" took more risks than banks that weren't.[30] We might expect the same type of behavior from defense contractors who don't have to worry about going overbudget, being late to deliver, or failing to reach their goals.

Precision-guided munitions may strike where they are programmed to hit, as when NATO bombed the Chinese embassy in Belgrade, but they are not going to discriminate beyond the set of coordinates guiding their flight.[31] The integration of intelligence, analysis, and targeting are not nearly as refined as vendors would have us believe. Missiles hit houses with families in them, and enemies put innocents in places where they will be targets. ABC News, in July 2014, showed Palestinians running around on the roof of a building that the news crews knew had Hamas facilities in it. They were going to be bombed. The Palestinians were open to endangering their lives to save the building because they knew the press would be actively reporting on civilian causalities. The quality of munitions and guidance systems can't overcome the willingness of people to put their lives at risk.

Stealth aircraft do not win wars, but they penetrate airspace when manned missions are necessary. They sit for all but a few hours of their lives doing nothing. High-priced hardware (usually overpriced, and overbudget before it ever gets to the field) is far too common. We are not helping the soldiers, airmen, and sailors who fight to do so better or less often.

There are many more companies producing war materials than banks that can't be allowed to go under. They are too big, or sometimes too exclusive, to national defense to be permitted to fail, so they are propped up. The Defense Department has a set of contractors that feed it weapons, supplies, and services. Defense maintains production of items that are not needed simply to keep the contractors in business and help them grow.

Suppose we had only one company that made flack vests for our army troops. The army doesn't need as many flack vests when it cuts troops, so it starts cutting back on purchases of the vests. The one company that makes them will thus have no work, unless it has opened up other markets and can sell the vests to other countries or different types of businesses, like police. But some of the items that are critical to the military have a specific use that can't be applied to other industries.

Do we let this company go out of business when we stop needing the large number of flack vests produced? No. We buy flack vests we don't need, using the justification that we don't want to lose the capability to make these vests. This is an application of the Keynesian economic model first popularized after World War II: when a segment of the economy can no longer sustain itself, the government purchases excess capacity to keep the segment viable. That seems to be OK for most people, especially the congressmen and workers in the district where flack vests are made. The same principle is applied in many segments of the industry. We still have the Abrams tanks and C5As to prove this point.

The inventories of many weapons are being reduced as the war in Afghanistan winds down, but money is spent on them so they can be upgraded before they are mothballed or scrapped. In some cases contractors have incentive contracts that allow payments to close down assembly lines. The cost to stop production can be as much as the cost of sustaining it. Each one of the expenditures is small, relatively, but they add up.

There was an interesting campaign that started in 2014 to save the Tomahawk missile, among other things, being cut back by the Defense Department.[32] The characterization of defense contractors on the edge of extinction,

12. The Integration of Business and War

desperately trying to "preserve a capability" in the war machine infrastructure, was exactly the kind of propaganda the defense industries love.

The navy proposed ending production of the new Raytheon Tomahawk in 2017. You only have to see one of these devices flying to its target to appreciate the appeal: It doesn't have a pilot. It follows a path laid out for it, straight to the target, flying low and slow. When it gets there, it has enough explosive power to destroy a bunker, a building, or communications links. If we only looked at their advertising, we would think they were invaluable.

When the navy said it was time to end production, a hue and cry filled the halls of Congress warning that closing down the production line would do lasting damage to the industrial supply chain for these important missiles. They are using the same techniques used in information war. The navy has 3,000 Tomahawks, which can sit on the shelf for 30 years. Only 2,300 have been fired in combat, giving reason to believe that the navy won't need many more for the time being. In 2019, the navy plans to upgrade the ones it has with new capabilities, so the missiles won't be thrown away. Of course, it is a long time until 2019.

Raytheon says at least a dozen companies would have to close or convert to other production if the current level of manufacturing were to stop. The Pentagon Program Manager says the navy recognizes the importance of the supply chain and is looking for alternative suppliers. This is in contrast to Congress, which has said it will help the suppliers of things like Bradley fighting vehicles and Abrams tanks by keeping up production.

These weapons are embedded in the total cost of war with Iraq and Afghanistan since 9/11, which the Congressional Budget Office estimates to be over a trillion dollars:

> Assuming an annual level of the current Continuing Resolution (H.J.Res. 44/P.L. 112-4) and based on DOD, State Department/USAID, and Department of Veterans Administration budget submissions, the cumulative total appropriated from the 9/11 for those war operations, diplomatic operations, and medical care for Iraq and Afghan war veterans is $1.283 trillion including:
> - $806 billion for Iraq;
> - $444 billion for Afghanistan;
> - $29 billion for enhanced security; and
> - $6 billion unallocated[33]

Perhaps most of us did not even know we were taking care of the injured warriors of Iraq and Afghanistan, or part of the pension fund for defense contractors, and lumping that into the defense budget. Money is hidden away in many areas that are debatable, but never debated. Defense Departments don't want it being discussed, and neither do the defense contractors, who do their best to keep these kinds of details out of the press. This is a worldwide phenomenon, not one unique to the United States.

This is one of the most confusing research areas, because it is difficult to define when something is a war material and when it is a commercial product. A prime example is the Toyota pickup truck.

Almost any insurgency in the world has a certain number of photos taken with rebels riding around in trucks that are completely open and unprotected. They mostly seem to be Toyotas. At one time the CIA had one of the largest fleets of Toyota trucks delivering supplies to Afghanistan, where they were then transferred to mules to move into the interior.[34] Toyota trucks are reliable and cheap to maintain, compared to an armored multi-purpose vehicle or a mine-resistant armored vehicle, and faster than a horse or camel. Nobody is going to try to stop Toyota from exporting trucks. Toyota might even take pleasure in the fact that its trucks can go anywhere and survive (at least until a Hellfire leaps off a Predator and blows the trucks up).

War materials are the fuel of war, but there are significant differences in the quality of the fuel. China is one of the largest arms manufacturers in the world, and one of the most prolific arms sellers to places where there is trouble.[35] China does not export as many arms as some others, but it covertly exports to countries where it may stir up fighting—places like Algeria, Angola, Bangladesh, the Democratic Republic of the Congo, Guinea, Egypt, Indonesia, Iran, Iraq, Jordan, Kenya, Libya, Myanmar, Pakistan, Sri Lanka, Sudan and Zimbabwe.[36] We could sell fighter jets to all of these countries without much impact on which side wins the battles. Jets are important, but we have learned from regional wars that they don't make a whole lot of difference to the outcome. The Russians had the HIND helicopter, the SU-25, and MiG-21, but they could not overcome a force of fighters armed with very few anti-aircraft weapons. Nor could a coalition of world powers that came after.

China is also one of the largest customers of Russian advanced weapons, and the second largest importer of weapons overall.[37] We almost never hear of direct arms sales by the Chinese because they are more careful than most about how and where they sell them. China sells weapons covertly. The shoulder-fired anti-aircraft missile known as the FN-6 is an example of sales that are not obvious and do not get recorded. China still has become the fifth largest exporter in the world, even though the Chinese hide a good bit of their transactions. If we could discover all of their transactions, they could be number two.

When the rebels in Syria wanted arms and the West wouldn't supply them, the Sudan did.[38] On the surface, that wouldn't seem to involve China. But the weapons were made in Sudanese factories and China (among them the FN-6, which is a dangerous weapon for helicopters and small airplanes that had provided an advantage over the rebels). These new weapons were paid for by, and often routed through, Qatar.

In July 2014 China sold $38M worth of missiles, grenade launchers, and machine guns made by China North Industries Group Corp to South Sudan, promising to facilitate peace between the North and South.[39] But the Sudan already made its own weapons. In 1993 Sudan started constructing them under the banner of the Military Industry Corporation, which is partially supported directly by Iran, and partly by Russia. The Sudanese make versions of Chinese weapons, and both China and Iran have sent advisors to help them. A report by Radio Dabanga on arms manufacturing in South Sudan says:

> A technical review of Sudanese manufactured weapons confirms that they derive from Chinese, Iranian, and Soviet designs. It is not clear whether Sudan simply repackages Chinese ammunition, or assembles cartridges that have already been marked by the Chinese. Because of Sudan's close military ties with China and Iran, it is likely that technology for the production of weapons and ammunition was supplied from the two countries. It is unclear whether any formal licensing agreements exist.[40]

Qatar buys arms made in China and Sudan for Syrian rebels who are favored by Russia, China, and a third of the countries in the Middle East. Nobody can prove that China had anything to do with the sale. If we look for troublemakers, we have to list China among them, with the Russians and Iranians close behind the United States.

If Lockheed sells an F-35 fighter to Yemen for $159M, it gets a big number posted on its sales figure, but the cost of the aircraft is not the total out-of-pocket cost. F-35s are big and difficult to hide when on the ground, and they require an extensive support structure that is far from being as stealthy as the aircraft. Yemen gets a great conversation piece, but one aircraft is not going to help very much. As part of the "package" it will need a place to store and repair the aircraft, a lengthened runway or two, and training for ground crews and pilots. The add-ons account for most of the difference between the FMS figure, the booked price of the aircraft, and the total cost of a package that a country purchases.

If China sells Yemeni fighters AK-47s for $250, which may include a small profit for those making them, it is very economical for the fighter who would otherwise have to buy them on the black market for a few thousand dollars. For $156M China can arm half the African continent, and the people they arm are grateful the weapons are inexpensive and easy to acquire. Everyone knows war is expensive, but there are ways to reduce the cost of acquisition and total cost of ownership. Selling cheap weapons is one method, but the best and most cost effective is cyberwar.

The Cyber Arms Merchants

In the 1970s and 1980s the arms merchants got into an even more dangerous game. They described this as work in cyberspace, the next great battleground.

There is almost no company that makes weapons systems that says it is not expert at defending or attacking somebody's computer networks; yet most of them have been successfully hacked themselves, and so have their clients.

These companies don't want to talk about the intrusions; they won't confirm or deny the reports, because getting hacked when one is supposed to be defending others from getting hacked is not good for business. Reuters reported that several defense contractors were hacked after the Chinese stole RSA secure token software.[41] Krebs also documented a report by Cyber Engineering Services, Inc., that indicated three companies in Israel—Elisra Group, Israel Aerospace Industries, and Rafael Advanced Defense Systems—were hacked in 2011 and 2012 from servers in China. The material the hackers were looking for was related primarily to the Iron Dome missile defense system.[42]

In the United States not one weapon systems contractor admitted losing information, but the Defense Science Board published a list of them within the U.S. government, so we know who they are anyway. It is hard to keep a secret in Washington.

Government agencies that hire contractors to design and build secure network components find that these supposedly secure network components do not protect them from attacks. Either they are taking the wrong approach in what they ask contractors to do or the contractors cannot do what they are being asked. Neither result is satisfactory.

The Russian Business Network probably understands hacking as a business better than most industries, but there is always room for competition in such a lucrative field. Enter the world's big defense contractors and federally funded research and development contractors. (The 40 FFRDCs are quasi-government organizations that can act in many capacities as government officials, although they are not. Between 2008 and 2012 they were paid in excess of $84 billion.[43]) These companies pretend to have thousands of experts who can defend or attack any network in the world. These are the White Hat hackers—the good guys.

At a national information warfare conference, I was sitting next to a person from a U.S. ally, one of the five-eyes countries. We were getting a series of "capabilities briefings" from several companies, and one of them said his company had 400 representatives who were specialists in information warfare across the range of cyberwar disciplines. The guy next to me almost choked on his drink, and I thought he might have gotten some coffee down his windpipe. I asked him if he was OK, and he said, "My God, we don't have 400 IW specialists in our whole government." Their presence is government should infer that government agencies would be able to protect themselves from external threats and protect government information, when that doesn't seem to be the case.

In the 2014 budget request most areas of the Defense Department were cut, but one area that was not was cyber. We might ask ourselves why, if the military is involved less and less in the elements of cyberwar, its cyber expen-

12. The Integration of Business and War   157

ditures are staying the same, or even increasing. Defense contractors have locked onto the cyber revenue stream. That by itself is of no particular concern, but in cyberwar the defense contractors have become combatants when they shouldn't be.

What most arms merchants sell, where we can see it, is cyber defense, which is common in both the business and the government sectors. Cyber defense is defense of computer networks against attacks from outside. The *New York Times* could have used some cyber defense when the Chinese attacked its system. Defense keeps cyber thieves out. At least, that is what it is supposed to do.

In defense, we have security engineers and architects who are designing and building secure systems, and security specialists in intrusion detection, incident analysis, and emergency response. These are expensive resources to obtain, keep current, and keep billable at the rates they require to be profitable. When they are good, like the team I had at EDS, they never have to look for work; it finds them.

The losses from cyberattacks fall into several categories of damage to the victims. Companies, sometimes out of fear, try to find the most cost-effective solution to mitigate further risk. Most of the victims don't know the exact potential damage they are exposed to and vendors exaggerate the risks (to their benefit). Most definitions of potential risk are estimates, almost always favoring the companies that provide the solution, so they are suspect from the beginning. Such definitions are more accurate when discussing the reasons for concern about defense.[44]

What defense contractors also sell, without talking about it much, are Computer Network Attack (CNA) and Computer Network Exploitation (CNE). In November 2012 the President signed top-secret Presidential Policy Directive 20, "U.S. Cyber Operations Policy," which slightly changed the use of CNA and CNE and appeared *in toto* in the *Guardian* newspaper. (We have to ask what the point of having a top-secret policy is when it is disclosed to the press.) CNA is now called Offensive Cyber Effects Operations (OCEO), which includes various aspects of both CNA and CNE.

> Presidential Policy Directive 20 defines OCEO as "operations and related programs or activities ... conducted by or on behalf of the United States Government, in or through cyberspace, that are intended to enable or produce cyber effects outside United States government networks."[45]

CNA and CNE are not new, but combining them is. CNA and CNE were isolated parts of the same general skill set, intentionally separated from network defense. That separation exists now in definition only. The policy allows "several departments, including the department of defense ... to conduct domestic operations without presidential approval."[46] This essentially allows the Defense Department to conduct operations on domestic networks, when the DoD was originally confined to protection of its own networks as a precondition of Congress in setting up the so-called Cyber Command. Cyber Command is under the control of the Strategic Command, under the Joint Chiefs of Staff. In cyber, the military will continue to be part of the fighting, and will continue expand its scope.

Within the federal government CNA and CNE have been looked upon as "inherently government functions," without anyone knowing what that really meant. There has always been a question regarding whether any contractor should be allowed to perform any function that involved CNE or CNA. Remember that Edward Snowden had access to many of these areas and published documents related to each. That creates a question about whether he was in a position that was inherently governmental, and what it means if he was. Contractors deal with these issues every day, but favor a loose interpretation of that term.

A definition will not help very much. The policy is contained in Public Law 100–417, and described in greater detail in the Office of Federal Procurement Policy, OMB 11–01, dated October 12, 2011. The policy is not very clear. This is guidance on when services are outsourced to contractors and when not, but it is a mess:

> (c) The following is a [shortened version of the] very long list of examples of functions considered to be inherently governmental functions. This list is not all inclusive, and is far too long, but it shows the intent of the legislation—to describe what is an inherently government function:
> (1) The direct conduct of criminal investigations.
> (2) The control of prosecutions and performance of adjudicatory functions other than those relating to arbitration or other methods of alternative dispute resolution.
> (3) The command of military forces, especially the leadership of military personnel who are members of the combat, combat support, or combat service support role.
> (4) The conduct of foreign relations and the determination of foreign policy.
> (5) The determination of agency policy, such as determining the content and application of regulations, among other things.
> (6) The determination of Federal program priorities for budget requests.
> (7) The direction and control of Federal employees.
> (8) The direction and control of intelligence and counter-intelligence operations.
> (9) The selection or non-selection of individuals for Federal Government employment, including the interviewing of individuals for employment.
> (10) The approval of position descriptions and performance standards for Federal employees.

12. The Integration of Business and War

(11) The determination of what Government property is to be disposed of and on what terms (although an agency may give contractors authority to dispose of property at prices within specified ranges and subject to other reasonable conditions deemed appropriate by the agency)....

(14) The conduct of administrative hearings to determine the eligibility of any person for a security clearance, or involving actions that affect matters of personal reputation or eligibility to participate in Government programs.

(15) The approval of Federal licensing actions and inspections.

(16) The determination of budget policy, guidance, and strategy.

(17) The collection, control, and disbursement of fees, royalties, duties, fines, taxes, and other public funds, unless authorized by statute, such as *31 U.S.C. 952* (relating to private collection contractors) and *31 U.S.C. 3718* (relating to private attorney collection services)....

(19) The administration of public trusts.

(20) The drafting of Congressional testimony, responses to Congressional correspondence, or agency responses to audit reports from the Inspector General, the Government Accountability Office, or other Federal audit entity.

Logic would say this type of work should be done by a government employee, not a contractor. But logic doesn't hold here, and *inherently governmental* does not mean "performed by government employees"—it only means "performed under the supervision of a government entity." Snowden proved that this definition is obviously flawed because contractors were responsible for supervising the work of other contractors; the government supervised the contract, not the work being performed under the contract. Contractors perform almost every function listed above, making "inherently governmental" a meaningless term. There are times when we need to consider what the term *should* mean, and cyberwar is one of them. We need to ask ourselves if we really want contractors performing CNA or CNE functions in our government or, for that matter, any government.

We know the Chinese and Russian militaries do CNA and CNE because several attacks have been linked to their hacking. As explained earlier, in China and Russia the lines between government (including the military) and business are blurred. Even with the blurred lines, however, there are good reasons to question the idea of armed forces being in charge of cyberwar. The Chinese, who have problems with their military going its own way now and again, should not have their military hacking foreign systems; yet they do. Too often, the military objective is to build weapons that can compete with the country the Chinese are stealing designs from. These weapons are manufactured by companies that are still tacitly controlled by military units. In the case of cyber weapons, no country should want the military or a contractor in control. What is being said is that it is OK to have a contractor's employee peering into the networks of some other government, or contractor, in another country.

Contractors would argue that modern war is no different from fixing an engine for a combat aircraft, but reaching into another country's network *is* different. This is especially true in areas that involve potential combat roles to be played by an attack. A contractor should not be pulling the trigger on that kind of weapon, and shouldn't be a position that would allow them to do it. It should be a government employee, under government control. That is especially true in authoritarian regimes where the military units and governments seem to come from different planets, and not just different perspectives.

For most of the early years of information war, contractors and government employees were limited in what they were allowed to do. If a person did computer network defense, he was not allowed to move in and out of CNA or CNE. That separation has been lost because we have forgotten the original purpose.

As a practical matter, a person who defends networks has a good understanding of the kinds of attack methods used by adversaries. Conversely, a good CNE person has deep knowledge of how to get into networks and pull information out without being discovered. A logical person might say it would be a great idea to team these people up so they would both become better at what they do. But logic doesn't apply very well in this context. In professional football the offense seldom practices against its own defense except once or twice in the pre-season to get the fans excited about the team. If they practiced against each other throughout the season, the only defense the offense would be really successful against would be its own. Instead, the teams have two "practice squads," whose players learn the next opponent's offense and defense and emulate them. The offense practices against the emulated defense of the next opponent, and the defense practices against the emulated offense. In the same way, except as an occasional special exercise, having CNA and CNE disciplines working together, or competing against each other, detracts from their missions of either protecting U.S. assets from foreign attacks or exploiting foreign assets. They interfere with each other.

There will never be "bulletproof" defenses for computers on the Internet, no matter how many Russian Business Networks might spring up. At one time, we thought that was possible, established criteria to measure it, and set standards to build very secure systems. That didn't last. Technology changed faster than the government's ability to keep up, and it is getting harder, not easier, to do. Hackers got better at finding ways to get in; defenders got better at stopping some of them. As long as the sides were in balance we could accept the losses that occurred, sometimes by insuring against them. They are not in balance any longer because defense is not valued the way it used to be.

12. The Integration of Business and War

The question then becomes, if we can't do both well, should we attack or defend?

Defensive and offensive computer operations are always at odds, and every country makes some kind of strategic decision about how those will be balanced. Good defense makes offense harder to do, because it tends to spread technologies all over the world to countries both good and bad. Countries soon find it harder to get into each other's networks.

All the major powers have demonstrated they would rather have the information from their intelligence agencies than the defense of their own networks. Not one major power, not China, Russia, Great Britain, or the United States, has good computer security in its computer networks, but all of them demonstrate an ability to extract information from other computers. (Canada and Australia are probably the best defenders, which is why we don't hear about them all that much.) That means the value of the information they get from other people's computer systems exceeds whatever costs come from not having good defenses. This is an issue that is hard to explain.

When we say there is always a conflict between defense of computer networks and what intelligence services do in collection of information, we are recognizing that no government can have it both ways. If a government has very secure networks, the means for making networks secure will spread all over the Internet. People would then get better at security of information, which, we would believe, is usually a good thing. Other companies and governments would use these methods to make collection of information by computer harder to do. The intelligence services then would have to work harder to get the same information. They would like for things to be easier, not harder.

In a recent case brought by privacy advocates in the United Kingdom, Charles Farr, director general of the Office for Security and Counter Terrorism in the UK Home Office, described how the offices might justify interception of e-mail without admitting that anything of the sort was being done.[47] If e-mail is between two British residents, even where they may go through equipment in another country, investigation would require a warrant that names a person or place as the target. Warrants for Google searches executed by British residents, however, are likely resident on foreign servers and do not have to name a specific person or place. The latter makes it possible to issue a broad warrant for "all persons searching" a particular website, like ones involving terrorists or criminals. The fact that information about searches and other uses of the Internet to acquire information is available to the government bothers privacy advocates, although it really shouldn't.

We may blame the NSA for collecting information from cell phones and whatever else it has managed to get into, but the world has not made the NSA's job harder by making the networks more secure from the kind of threat to personal privacy that the organization demonstrates. It was difficult to believe anyone could get the cell-phone traffic of a head of state, and we can't blame any intelligence function for doing it if it can be done. That is their job. There were probably other countries collecting the same data from the same cell phones, because that is the purpose of intelligence services. Switching from one carrier to another is not going to help the situation unless the government controls the one they are switching to. We are criticizing the NSA for doing what we pay its agents to do well. We should instead blame the companies that intelligence services take data from. Google and Apple, among others, have attempted to encrypt their networks internally, and in an untitled speech at the Brookings Institution FBI Director James B. Comey made a point of saying this kind of encryption is not good for national security. The commercial businesses can't win, but that doesn't mean they will stop encrypting.

Intelligence services may try to do things within their governments to make the job of gathering intelligence easier, like the direct connections Vodaphone experienced with certain countries. There is nothing "wrong" with doing what they have to do to collect information, nor with the measures they use to be successful. They just shouldn't get a blank check from the rest of the technology world. Advertisers, marketers, telemarketers, local businesses, and government agencies buy or collect information they then use to reach potential customers. Our phones ring all the time with some of these calls. We blame them for calling us, when we should be blaming the people who sell the information to them. Apple and Google are damned by the government if they do encryption, and damned by their customers if they don't. They are on the right track, making their systems more secure.

This is out of balance. The world's intelligence services, particularly those of China, Russia, France, Israel, and the United States collectively, are winning out over other functions of business and government. Some will use it to augment their business capabilities for Economic Warfare; some will not. Some will use it to deny civil liberties to their own citizens, and others will not. But for balance to be restored, defenders have to be encouraged to do better. We have to put more money into the defense of networks, the defense of information, and we need to organize business elements that want to be helped. Part of restoring balance is raising awareness among users to demand better protection of networks from our carriers and their big customers, but they can't demand it in one country without demanding it in all countries.

Neither governments nor users have been strong enough to get service providers to improve network security either by mandate or by paying for improvements. Our own government is not capable of doing that because it is dominated by industries that don't want mandates for tighter controls

12. The Integration of Business and War

placed on networks. Congress is unable to pass legislation to allow business entities to share information without liability for mistakes. The industry says it is afraid of anti-trust enforcement if the legislation passes, a novel and unique way of showing the companies' disinterest in anything that might allow them to be liable for the software and hardware they put on the Internet. As long as our businesses don't show a greater interest in making their systems secure, governments will be reluctant to do much more. The intelligence services will continue to dominate. And they will have plenty of help from their contractors.

13

THE COMBATANTS

Cyberwar is far from new, having been well defined in the 1980s. However, it has changed a lot since then. Modern cyberwar takes its cue from the idea that computers can be networked in clouds that are owned by multinational corporations and governments. They are integrated into networks that interconnect other clouds with corporate and government computers. We no longer control our own information, and some governments take advantage of that. These are wars over information, no matter where it is held or who controls it. The people who fight are a peculiar breed of specialists who know how to get to it, wherever it is.

I used to sit near a woman who was one of the best hackers in the world. She sat with other elite hackers, and she was smart, quiet, straightforward, and focused. I said, "Good Morning," to her every day for three weeks before she finally replied. She wasn't unfriendly—just focused. Nothing existed for her but that computer; for people like her, such focus is not that uncommon.

Every hacker has an ego that comes from roaming through someone else's network without being caught. Most are rebels in some way. As a consequence, they are reluctant to ask for help. They share information, but not all information they have. It is very much like having a trade secret; each hacker has their own techniques and tools, and they are often only willing to share information they presume that anyone could know or easily find.

Hackers believe they can overcome any obstacle through persistence, because there are enough users and developers out there who make mistakes that will let them in. The vast majority of the time they will be proven right.

Over time I noticed more of the workers coming to this woman. They would sit and chat about topics related to their work, and she would discuss the problem. Then she would ask them a few questions about how they were approaching a solution, using what they knew and what they did not know. It usually took less than a minute or two. She would offer a suggestion and they would walk away. She did this a couple of times every day. In a few months her involvement slowed but never stopped. The office never treated her any different from the rest, but the hackers treated her with respect. Nobody outside work ever knew what she did, where she did it, or that she was better at it than any of her peers. Hackers, at least the good ones, are anonymous in their virtual world and choose anonymity in the physical world as well.

There are different motivations for hackers. There are out-and-out criminals who steal for profit. For a few it is simply a game—they seek no reward other than the satisfaction of knowing that they beat a corporate entity like Microsoft, Sony, Symantec, or Oracle. Others want to impress their friends and family. This group is more annoying than dangerous, but they unfortunately develop techniques and expose vulnerabilities that can be leveraged by those more sinister. A few of the more gifted hackers are hired by criminals and governments, which lack the technical expertise to get into certain types of systems. There are also political activists and anarchists. Criminals, activists, and anarchists are not very popular with governments on any side of war.

Anarchists like Jeremy Hammond are the exception. He clearly stated his political ambitions as anarchist-communist, and at a 2004 DefCon conference he said a few things about bombing the Republican National Convention that the FBI chose to discuss further in person.[1] He worked with Anonymous, and against those who tried to discover who and where the members of this group are. His opponents included, in his guilty plea, the FBI, the Arizona Department of Public Safety, the Boston Police Patrolmen's Association, and the Jefferson County, Alabama, Sheriff's Office. He was indicted for stealing records and credit card information from Strategic Forecasting, Inc., a private intelligence company, and then releasing some of the records to the public. Four other members of Anonymous were indicted for separate offenses. In response, Anonymous released personal information about the judge in the case and her husband.[2] Hammond had previously hacked HBGary Federal, a security firm that employed a person who threatened to identify and expose the members of Anonymous.

In an RSA-commissioned study of hacker traits at Danube-University in Krems, Austria, an attempt was made to determine the psychological characteristics of people who steal identities.[3] We could summarize these findings in a few simple statements: Hackers are largely male, liberal with libertarian traits, generally not religious, highly intelligent, and bad at interpersonal communications, and they have "a facility for intellectual abstraction." Bernardt Lieberman, who interviewed 42 hackers during his career at the University of Pittsburgh, found their motivation was understanding computing, not the thrill of getting information from someone. While there may always be an element of that thirst for understanding in every hacker, money drives the vast majority.

When people use the term hacker today, they are usually talking about criminals and those who work for or contract with governments to do hacking for a living. Some have serious jobs with organizations like the Secret Service, the intelligence agencies, and the equivalent national police force, like the U.S. FBI. These people have to prove in court that someone actually did the things they are accused of, by collecting and simplifying evidence. Simplification can be very difficult because juries and judges usually do not have in-depth technical knowledge—certainly less technical knowledge than the people providing the evidence.

There are not very many hackers who are really good. If a person attends a conference at a place like Black Hat, they are prone to believe that the world is full of hackers who can do what those guys on TV do. Most attacks are not original; they are copied from what has worked for someone else. Less than one in a thousand develop new techniques. The rest are learning from attacks developed by others. In many respects they are like scientists. A few are at the edge of their technology, pushing it forward; most are using the science to earn a living.

The same is true of defenders of networks. They don't have technical knowledge of more than a few network component types, attack techniques, and balanced defense methods that allow efficient operation of the network but still provide good security. They have certificates to prove they have some knowledge, but no specific skills. There is no reason to blame them for that. Hiring authorities are not good at understanding the skills required.

Governments focus on process rather than technical skill, which degrades the overall career field. Hiring has become a rote activity, driven by hiring managers who don't know the field at all and are required by policy to select people with certifications of dubious value in the work that has to be performed. Most certifications are general knowledge tests, while the skill of attacking and defending networks is very specific to hardware and software types. The difference in approach is why an organization like SANS is so popular with professionals. At its core, thanks to Stephen Northcutt, SANS teaches performance-related skills, not educational concepts.

Good attackers and defenders are difficult to attract to government work. Their skills are undervalued, the process orientation is stifling, and the policy is voluminous, inconsistent, and unenforceable. In the United States, Homeland Security recently attempted to establish hiring incentives for some good technical defenders, but it has yet to get that measure through Congress. Only a few places in government hire, train, and provide a career path for technical specialists in CNA, CNE, and CND. The vast majority of agencies do not.

Sifting through networks anywhere is not a simple thing to do. The average computer user never has to do that because connecting to a service provider has been made simple to do. If you buy service from AT&T, it takes care of connecting to all the sites you might want to visit. If you are doing research on Turkey, it is easy to connect to websites in countries in and around Turkey, with the same ease that you could connect to a home network. The content will be in another language, with different kinds of control over the networks, but the connections are there. The hardware and software is fundamentally the same; Windows, Unix, and Android/Linux are the same everywhere in the world, and the hardware platforms may be older or slightly different, but computer architecture and instruction sets have not changed dramatically in the past 10 years.

Hackers are usually not interested in connecting as a user, because that limits what the system will allow them to do. They want to have administrative access—that is, privileged access. That takes time and patience.

Administrators (admins), like Edward Snowden, are a privileged class of people on the Internet with a high level of trust. They are capable of commanding a computer to create users, give access to another network or person, shut down someone's access, update the network components, and manage the log and audit records of a system. What hackers want to do is collect administrative access to as many computers as possible, so they use some interesting techniques to do that.

The simplest way to become an administrator is to steal log-on credentials, usually by asking for them, which is called social engineering. A couple of websites advertise variations on this theme and tell a hacker to pretend a virus is invading the network and the administrator's credentials are required to help stop it. It sounds preposterous that any administrator would fall for this kind of trick, but they do.

There are many more ways to do the same thing, by taking advantage of holes in the way administrator accounts are managed inside each operating system. These hacks are posted on the Internet; some of them require a program running on the network of the system being attacked. In one case, this was accomplished by leaving flash drives with files hidden on them that extracted and loaded when the drives were inserted by people in the target building. The parking lot of the target building was the distribution point for the flash drives.

In October 2012, Israel was attacked using a RAT called Poison Ivy, causing the Israelis to ban the use of USB devices and cutting off access to the Internet for all police. These attackers were found to be going after both Israeli and Palestinian targets, as well as the United Kingdom and the United States.[4] These kinds of attacks are public knowledge, and they become more difficult as vendors of equipment become more sophisticated at countering them, and attackers change attack vectors to challenge defenders. What we should fear more than this is an insider, a trusted person, gone bad.

In May 2013, Edward Snowden entered a hotel in Hong Kong carrying four computers, at least one of which had national security information on it. This was not the run-of-the-mill secret content that WikiLeaks scattered all over the Internet, but a special kind of Top Secret (TS) called "Sensitive Intelligence (SI)."[5] The WikiLeaks material was tame by comparison. Few people have ever heard of TS/SI; yet, judging from the traffic on the *Guardian* website, lots of them got to see it. A few people might even have had security clearances, polygraphs, or computers with access to the top-secret networks that Snowden had, but it would have been only a few. The rest would have had no clearance to see anything like it—yet they could.

Snowden was a paid employee of a contractor working for the NSA. His worth to his friends and enemies was based on the perceived value of the information he had in his possession, and not his sparkling personality and wit.

He also stepped into something else: a war waged by the United States and its allies against China and its best friends (i.e., Russia, Iran and North Korea). The United States and China are at the core. Both of them have been clever about how they conduct this war and both have been careful to avoid the term *war*, but that's what it is. It is a secret war, about secrets—how they are used and how they are collected—held by different members of the alliances. Snowden represents a new round in this conflict, and the evidence indicates he was a Russian asset, from which the Chinese will benefit. It will be a long time before we know much about how Edward Snowden came to defect from his country and try to find a home somewhere else in the world. In the meantime, he chucks out pieces of classified information for all on the Internet to see. The leaked information that does the most damage has nothing to do with him.

In the middle of January 2014, Mike Rogers from the House Permanent Select Committee on Intelligence and Homeland Security Committee Chairman Michael McCaul were both on multiple talk shows saying that Edward Snowden was a spy and probably had help from the FSB, the Russian secret police. The KGB, predecessor to the FSB, was Putin's home before he became president. The news media called this a "revelation," which only meant they hadn't thought about it before these two stated the obvious.

In August 2014, *Wired* published an article by James Bamford on his interview with Snowden in Moscow.[6] Bamford is no stranger to the NSA, with a long history of writing about what its agents do and how they go about it. He is certainly not their friend. He used every cliché ever associated with Snowden's defection, including his regrets at being part of a spying operation that hacked places in China, elaborating on similar Chinese claims on the same topic. Bamford said others provided documents that Snowden did not give to the press, a convenient misdirection allowing the release of embarrassing

13. The Combatants

information without attracting attention to the true source. It sounds like something the Russians and Chinese would want to have written, with Snowden exactly following their script.

Snowden said he was distraught over things he saw the government do and wanted to expose "abuses" that ran contrary to what a good government should do. These were the same words that defectors who came before him used. But he didn't act like a fellow who saw something on his computer and defected because he wanted to help the United States be a better country. Rogers said Snowden looked only for things that could undermine the intelligence collection capabilities of the United States. He also said it looked like the Russians were helping Snowden, or at least that was being investigated as a possibility.[7] The Russians deny it.

In parts of the world, publishing this kind of information through the press would lead to the arrest of the people who did it. In the United Kingdom and the United States, however, the press is allowed to publish what it comes by. It does this in the name of freedom. Both the Russians and the Chinese have been very good at using the laws of countries they target to their benefit. Patent law, criminal law, privacy legislation, and telecommunications law are all helpful to them.

Edward Snowden was, in a way, a unique spy, but his revelations were in an area of spying that many people do not understand. He was a privileged, trusted person who used the trust he had, by virtue of his position, to his benefit. All governments accept a certain amount of risk in employing contractors in positions of trust.

What is not unique about Snowden was his insider status, something other good spies have all had. As did many before him, he collected and gave to our enemies the sources and methods used by the United States to collect intelligence, mostly by using computers. Some of that collection, especially domestic telephone records, was at issue, but most wasn't. He published the names and associated methods for many programs that were classified as top secret because they would cause exceptionally grave damage to the United States if disclosed. Almost everything he stole was related to how the National Security Agency collected raw information to analyze and craft into intelligence reports.

Edward Snowden compromised a number of intelligence operations. He has been identified as the source of many stories about the capabilities of U.S. intelligence, including sources and methods, some of its best-protected information. We will never know if Snowden actually released, or stole, half of the information attributed to him, but the perception is that it all came from him. The advantage to the press is they can use him as the source and nobody will ever know for sure if it was Snowden or someone else. The U.S. government avoids commenting on the accuracy or completeness of classified information released to the public, so we will never know if anything he said was accurate.

This wouldn't be the first time the Russians have helped a U.S. citizen steal classified information, but it was the first time the information they stole was made public. In 1960, William Martin and Bernon Mitchell, who said they had discovered that the NSA was listening to and reading messages of some of the U.S. allies, defected to what was then the Soviet Union.[8] At a press conference in September, Martin and Mitchell said they had defected as a result of their objections to U.S. intelligence methods, including the interception and decryption of the communications of U.S. allies. They went on to say that they had elected to come to the USSR because their own values seemed to be shared by a greater number of people there. In addition, they thought that the higher status enjoyed by women in the Soviet Union would make Soviet women more desirable as mates. That must have been an interesting press conference.

Our preoccupation with the NSA's business continues, in spite of only occasional flashes of it in the news. Critics of the NSA, and general intelligence collection, are not going to stop NSA from doing its job. We are not the only ones who value this kind of information. The Russians seem adept at getting it from our government employees and contractors.

The most interesting spy we never heard of was George Koval, who until 2007 was almost unknown outside the walls of the KGB. Koval was a spy for Russia during the Manhattan Project, the making of the first atomic bomb, and he would still be unknown had it not been for a slip-up by Vladimir Putin. Putin was at an award ceremony held for officers of the KGB, his old employer. Seeing Koval's picture on the wall, Putin asked who it was. That little slip led to the disclosure of Koval's name and his real work in helping Russia build a nuclear weapon, right after World War II.[9] While the United States was looking for Russian spies in every aspect of life, it turns out there were some in places the United States would never expect to find one. That is where good spies always are.

Until Snowden, John Walker (recently deceased) and Aldrich Ames were the worst, from the perspective of the United States, having betrayed some of our most sensitive secrets. Ames was in the section of the CIA that gave him access to the names of people who were spying for the United States in other countries, including Russia. Walker served in the U.S. Navy and sold to the Russians the cryptography keys that were used to encrypt hundreds of thousands of exchanges between military units. We find it hard now to comprehend the damage done by those losses in the context of their time; both men were part of a broader information war.

In 1967 Walker was a walk-in at the Soviet Embassy in Washington, a place watched constantly by U.S. intelligence services. Apparently they missed him that one time, because he managed to steal classified documents until 1985. Walker was not alone in his crime. He talked his son Michael into stealing documents from the naval vessel he served on. These documents were supposed to have been destroyed, but they were retained and passed to Walker. Walker also talked his best friend Jerry Whitworth into doing the same. His most damaging disclosures were the cryptographic key lists, which could be used to decrypt messages. When the North Koreans captured a spy ship called the *Pueblo* in January 1968, they flew 800 pounds of material to Moscow. After that the Russians had access to naval communications until the entire system was changed.[10] Walker retired from the navy and continued to operate his collection program until his ex-wife finally turned him in.

Aldrich Ames was by far the most damaging to the intelligence community, which relies on protection of its sources and methods to remain effective. Ames gave up both and, like Walker, was a walk-in at the Soviet Embassy in Washington. The Russians must have been amazed at their good fortune.

In 1986 the CIA realized that its agents targeting Russia were starting to turn up missing in greater numbers than usual. CIA leaders looked at the possibilities—a case of the codes being compromised like Walker had done, a bug in the CIA offices, or a mole in the agency. They looked at the last option first, partly because they had a list of people who knew who these agents were, which would help narrow down the field of suspects. There were only 190. They started looking at each person individually, but not until 1993 did they get the break they needed.

One member of the mole team got Ames's financial records and another put them together with meetings Ames had held with Soviet "sources," finding a correlation between the meetings and upticks in his bank account. They turned over what they had to the FBI, because the CIA does not have arrest powers. The following February the FBI arrested Ames. Ames had a list of people whom he knew had the same kind of access he did, and he was to use that as a distraction if someone got suspicious. However, the person he would have named was on the mole team carrying out the investigation.[11]

The other two most damaging agents who were selling secrets to the Russians were Robert Hanssen and Edward Snowden. Hanssen, an FBI agent, sold agents' names, and the agents subsequently ended up dead. Snowden exposed the sources and methods of some of the NSA's most interesting Computer Network Exploitation programs.

Robert Hanssen was working mostly in the area of counter-intelligence. He was supposed to be working to find spies, just as Ames was supposed to be managing them. Even though these men were walk-ins, the Russians know a good source when they have one. They promote and handle good ones very well. It is instructive that the most damaging spies in U.S. history were all spying for Russia, not China. The Russians must be doing something right.

Snowden represents a new way to spy, and we ought to admire the Russians for their novel and successful way of going about it. It is a smart way of covering their tracks while they steal everything they possibly can. The spy admits he is guilty, gets the press to cover him in the publication of the stolen information, and nobody even looks Russia's way for months. If someone from the FSB was helping Snowden, you can bet that person is not around for the investigation. His picture is on the wall of that Kremlin facility, along with Koval.

Where Snowden landed first was Hong Kong, which until 1997 was a British possession overrun by Japan in World War II, and then returned to China by a 1985 treaty. This is not enemy territory because Hong Kong has always been a middle ground between China and the rest of the business world. It is still tenuous about its relationships with China, and citizens protest relatively unmolested (at least until 2014) to show displeasure with how things are going there.

Snowden is connected to the publishers of his documents, the *Guardian* newspaper in the United Kingdom being one. Snowden has already given interviews to *The Guardian* and *Der Spiegel* that were specifically slanted for the audiences of those newspapers. His *Der Spiegel* interview identified Snowden's claim that the NSA and Germany were closely aligned and shared quite a bit of information that was similar to what the NSA was supposed to be collecting on U.S. citizens. The *Guardian* interview disclosed similar cooperation with Britain's GCHQ, the Government Communications Headquarters, one of the three UK Intelligence and Security Agencies, along with MI5 and the Secret Intelligence Service (MI6). It is the British equivalent of the NSA, which was said to be collecting the same type of data.[12] The interviews were tailored to the countries reading the newspapers. The Russians are clever about how they use the information they have managed to collect, because they have quite a bit of experience in this area.

We are fairly used to the idea that a member of the press gets to publish what he or she are given, even if that information would be damaging to the host country. We sometimes call this "freedom of the press," though that term may not accurately describe the ability to release potentially damaging information, whether the secrets are national security or the next new gadget that Apple will introduce at the Developer's Conference. U.S. and UK laws will not stop a newspaper from publishing this category of information even when we know it will tell our enemies and competitors things we don't want them to know. Even so, there was a small amount of reporting about GCHQ going to the *Guardian* offices in London to destroy hard disks where some of the information Snowden gave the *Guardian* was kept.[13]

13. The Combatants

The assumption that every government spies, but the United States spies more, comes out clearly in the aggregation of Snowden's documents being disclosed. That idea is probably wrong. The United States does not spy more; it spies better. This is also clearly the argument the Chinese made when the United States accused them of stealing us blind. If there is a difference between Chinese hackers stealing business information from companies all over the world and the NSA doing intelligence work for the U.S. government, the world press seems to have lost sight of the distinction. These disclosures were not a coincidence, although the general public would find it difficult to tell. All the public sees is a stream of information that looks and sounds like countries spying on their own populations, as well as every human being on the planet. There is a thin line between intelligence collection and theft of commercial secrets. It is a line often blurred by the press and government officials who benefit from having a public believe they are the same thing.

The Russians have used our own system of freedom and a free press against us, and the press outlets got Pulitzers for their work. Our laws can't stop the press from releasing anything Snowden gives them. While the President debates what metadata the NSA can keep under what circumstances, the Russians are laughing and celebrating. Until we stop them, they will continue to use an unbridled free press to publish our secrets. Until the U.S. general population is able to understand that we are at war and that these tactics are battles to win their hearts and minds, we will lose these battles. It is just good cyberwar, a war based on information.

Most of what Snowden disclosed was CNE, so it won't take too long for our enemies to start thinking about how to stop these exploitations from happening again. At the same time, our intelligence services have to start thinking about keeping another Snowden from doing similar damage and laying the groundwork to start over. It is a complicated part of the larger information war. Reconstituting intelligence programs will probably set the United States back years in its collection capability and its relations with allied governments, and nobody benefits more than the Russians and Chinese.

> One distinction between covert action and other overt activities, such as traditional diplomatic or military operations, is that U.S. officials could plausibly deny involvement in the activity. This "plausible deniability," however, is predicated upon the covert action remaining secret.

Snowden took care of a few programs by removing the veil of secrecy. Just the knowledge that they exist means they have less chance of being successful.

As spy services go, the Russians have had a string of successes that would be hard to rival. The CIA and FBI are enterprises the Russians love to get into, and their best agents walked in to their embassy. Their spies have gotten deep into the military and intelligence services to a degree no other spy service can match. They hack defense contractors and have done so since computer networks gave them access. We almost seem powerless to stop them.

14

FIGHTING A MODERN WAR

General Custer, at the Battle of Little Big Horn, was surrounded by Sioux and Cheyenne warriors and must have looked at those around him with regret. He should have been better prepared, and the realization of that came too late.

In a physical battle, the overwhelming firepower that we saw in Iraq leaves behind destruction and bodies lying beside the road. We can even bring along reporters to be sure it gets out to the public. We don't have enough of that kind of reporting in cyberwar, because there is not much to see. The covert nature of modern war won't allow that kind of coverage. It is impossible to see tampering in another country's elections, the damage done by the theft of intellectual property, or the economic losses stemming from manipulation of contracts. It is like erosion. It takes place over time and is not noticeable until that sinkhole opens up, and houses tumble down into the dark.

The will of the people to fight is not there. The most aroused we have seen Europe was when a civilian airliner was shot down in the Ukraine. Then sanctions against Russia seemed like a good idea, and additional ones were drafted in days. In response, the Russians moved more troops along the border, in what the Ukrainian government saw as a bluff. They seem to have been wrong about that assessment. The Russians have banned certain agricultural products from Europe and the United States, which seems to have hurt nobody.

It took months to arouse the world to the threat of ISIL, when ISIL has known about them for years. When these fighters left home and started to post videos of their atrocities, it became obvious they were worth going after. Anyone who moves fast—be it ISIL, Russia or China in the South China Sea—has an advantage over those who need coalitions and logistics. We are still not prepared to fight when the need arises, and we certainly are not prepared to win the will of the people.

When militaries were first documenting their thoughts on information war, and how it might change warfare, they had some ideas about how wars would be changed by battles over information. The Ukraine situation wasn't what they imagined.

In 1996, the RAND Corporation conducted a study for the Secretary of Defense, outlining what the new face of war would be.[1] RAND was asked to run exercises that would help define the key issues for policy development. When this first came out to the defense community, it caused some consternation over two issues: control of the Internet and control of the media. At that time, not many people had been thinking about those issues.

During that same period, the Ballistic Missile Defense Organization (BMDO), where I worked, carried out a series of studies with both the Nuclear Command and Control System Support Staff (NSS) and the Defense Threat Reduction Agency, to look at how information war might affect actual combat, particularly in missile operations. Our first two concerns were the inability of governments to separate any of their functions from the Internet (even the most sensitive national security systems) and the lack of aggressive monitoring of the networks owned by the government. From those studies came SHADOW, an advanced intrusion detection system that focused on finding and reacting to attacks in the time it took to launch an interceptor and find its target. We added a network mapping function later that identified the location and status of every component in close to real time. We were thinking about future attacks that would use methods we had not heard of yet, and these two things were a start.

What RAND and BMDO were doing in those years was looking at the opposite ends of cyberwar—the strategic end and the tactical network end. We certainly got some things right, but we were off in terms of scope. Cyberwar has jumped past our original thinking by getting into everyday life, stealing identities, information, and money as part of national strategies. It touches all of us now, and we have to think about it as personal, as well as a government strategy.

RAND was telling us things that should have led our leaders to think more clearly about what was coming. The cost of entry into cyber was not very great, and we were soon not going to be alone. That turned out to be right, but it has taken far longer than people might have thought in the 1990s. Iran, Syria, Egypt, Kuwait, Qatar, Saudi Arabia, the UAE, and the Sudan are all relative newcomers to this area, with only Iran showing any inclination to apply it broadly, or successfully. The United States, Israel, Canada, Great Britain, China and Russia have been into it for as long as anyone else.

Low cost should allow many more countries to be involved in cyberwar in some way. There are small groups that hack targets for profit, but these are not governments. Nigeria, Brazil and Romania probably have high rates of cybercrime, but there is not much evidence that their governments protect these gangs the way the Russians and Chinese do. The governments may not be capable of handling crime within their borders, but they don't willingly ignore it.

What this means is that our predecessors believed cyberwar was going to be easy, and that it would be possible for a significant number of countries to modify their capabilities to get into the new game. That does seem to have happened, but very slowly. More people may be involved, but governments are finding they don't have the technical sophistication or control to deal with the broad range of disciplines required to manage cyberwar. It is much harder than they thought, and much more complicated.

Perception Management

RAND identified one large area, perception management, as something we should look at more closely. Governments have become much more cognizant of how to manage perceptions of their conduct, though that can be used in a number of ways. For example, North Korea makes a graphic of a missile flying toward the United States and says it could carry a nuclear weapon. The North Koreans then overlay that with a video of a test launch of one of their long-range missiles. Analysts on television say it might be able to reach the United States. The vast majority of us think that is incredible, but we still think about it. My mother sees the image and panics. She believes they might do it tomorrow, and she still remembers those "duck and cover" movies in grade schools that told us we could survive a nuclear blast. The North Koreans are managing a perception that they are crazy enough to start a nuclear war, given time, but they can be persuaded (i.e., bribed) to put off that kind of thinking. That is a perception that has served them well in the past, and one that they perpetuate, in spite of how other people view them because of it.

The Chinese are better than that, and listen to what others say about them. When we were critical of them for having businesses owned and operated by their military, they publicly changed their policy. When we criticized them for having too many companies that were state owned, they recalculated how ownership by the state was determined. When we criticized them for stealing technology through U.S. subsidiaries of Chinese-owned companies, they brought attorneys to the United States to work those cases in U.S. courts. When the United States kept China from making purchases in its telecommunications infrastructure, the Chinese published policy to permit other counties to buy into theirs. (We shall see whether they actually allow it.) They manage perception of their own country both internally by controlling what people say and externally by dealing with issues that determine how other countries perceive them.

The Russians use less sophisticated methods, but they are still successful. Robert Gates, who was a Russian analyst long before he became Secretary of Defense, points to the fall of the Soviet Union as a defining moment for Russia. It was embarrassing. Putin thus feels a need to protect Russian speakers: "It's where he feels that Russians are being persecuted or where there is a prejudice against them, or ... there are laws or rules that disadvantage them economically, politically, whatever, that he will, at a minimum, have his intelligence officers in there working the problem. And in the case of Ukraine, and particularly Estonia and Latvia, I think squeezing them economically."[2] The Russians are dealing with an inferiority complex, brought on by the collapse of their empire. It makes them impatient. They may have let that affect how they conduct themselves.

When it comes to public perceptions about a country, Russia is falling fast in world opinion polls. Europe and the Middle East both tend to have majorities that view Russia unfavorably.[3] In the United States, 75 percent believe Russia is an adversary or "a serious problem."[4] The world has mixed opinions about Moscow, but majorities of people in other countries have a negative view. The perception the Russians haven't managed very well is that they are run by former intelligence agency personnel and organized crime figures. Putin also acts like a former KGB officer when he is supposed to be a head of state, which doesn't help to change that view.

The United States and its allies are probably better than Russia and China at perception management, with the United States far and away better. A Gallup poll found that there were 650 million people worldwide who said they wanted to move from their country permanently. A hundred and fifty million of those wanted to come to the United States, and the next three countries combined (the United Kingdom, Canada, and France) had only a hundred and thirty-two million.[5] Relocation is a powerful indicator of feelings for a country. We are talking about people who are willing, for many reasons, to leave their homes and make a new life somewhere else. That requires commitment and faith. We don't find large percentages of people saying they want to move to Russia (6 million, or 1 percent of the total) or China (less than 1 percent surveyed).

Half of all Chinese surveyed have favorable views of the United States, but 96 percent have favorable views of their own country. Russians had similar attitudes toward the United States for five years prior to 2014, when approval dropped to 23 percent, while 92 percent had favorable views of Russia. We can directly attribute that to Chinese and Russian efforts to manage what their populations say and see; it is proof of their success.

The U.S. population actually says it has 87 percent favorable opinions of its own country. If we look at the will of the people as measured by favorability, Russia and China do well at managing internal perceptions. The

United States has less favorable views of itself, yet more favorable views among people worldwide. It also spends less of its resources trying to maintain that perception. Cyberwar is partly about perceptions other people have of us. In that area, nobody can beat the United States.

Internet Containment

Ten years ago, RAND cautioned leaders about the lack of support for the government to "seize control" of the media, or the Internet, in the event of war. Today, it may be more practical to have a national policy to contain the Internet without trying to control it. That is the direction the Russians and Chinese have taken, and it seems to make sense.

In our examination of tactical networks, my colleagues and I were often surprised by the connections to the Internet that seemed to be made by contractors and government employees. There were usually policies forbidding such connections, but they did it anyway, sometimes against the interests of the people paying them. The lack of attention and focus by these well-trained people was inexcusable. Countries do not do enough to make sure there are no Internet connections to systems that are part of their critical infrastructures.

By policy, there are some functions of government that don't use the Internet, because people who manage them realize the benefits of isolation. The control of nuclear missiles, the National Command Authorities (which maintain communications for the senior military leadership), national security components, and certain intelligence functions shouldn't touch the Internet. Some of these functions have to use special networks, hardened against the effects of nuclear weapons.

The Internet owes its existence to an understanding of communications in nuclear war. At the time it was being developed, the main concern was the firing of nuclear weapons by both the Soviet Union and the United States. Remember those 2,000 weapons in the Ukraine. Our nuclear strategy was "mutually assured destruction," a term that has a nice ring to it—at least until we start to think about what it really means. With large parts of the communications infrastructure gone, communications had to be routed around the holes. There was no way to do that at the time, and the research came up with one: the packet-switched network.

Before we understood much about nuclear weapons, we used them on missile defense systems like the Nike series missile, which was described as "nuclear capable" without ever mentioning that one fifth of the nuclear arsenal was out in the suburbs around major cities.[6] We would cringe at the thought of it, 40 years later. What we also didn't know was that an electromagnetic pulse (EMP) generated when a nuclear weapon goes off knocks out a good bit of electronic equipment. Our weapons might have shot down those Russian bombers, but they would also have done a good deal more harm than the bombers would have. The Internet would not be working, and it wouldn't be the only thing.

An EMP will "disrupt satellite-to-satellite, satellite to aircraft, and satellite to ground communications" and "affect communications systems and power grids as much as a thousand miles" away.[7] In this situation, we won't be getting our news from television or computer, but it won't matter because we won't have electricity either. So, if someone asks why a nuclear country would be concerned about one other country getting a nuclear weapon, the power of that one weapon is the reason. It makes those threats by North Korea seem a little more real, and worth taking seriously.

We don't hear about nuclear war anymore; yet many countries still have enough nuclear weapons to annihilate the rest of the world. Russia and China have both adopted, and sold to their allies, a new threshold for war that is more about perception than anything else. Cyberwar is a better, less threatening threshold that keeps other countries from rushing to put money into the threat it represents.

With the convergence of networks and the rise of the Internet of Things, we are putting all of our eggs in one electronic basket. When someone takes it away, the consequences may be more severe than we realize. It is easy to see when nuclear weapons are being used, but not so easy with cyberwar. The first goal of cyberwar is to prevent the perception that we are at war.

When I said, at the beginning of this book, that we are involved in a war that we are losing, it is because we don't see it as war just yet, and by the time we do, there won't be much we can do about it. It will be over. Crimea was only one small example of how it can be done.

Control of the World's Networks

On September 11, 2001, an attack none of us will ever forget caused a lot of damage to telecommunications in New York. What gave the United States the ability to restore service quickly was Verizon's dominance in the New York market.[8] That kind of service dominance is gone in most places, traded for competition that reduced the price for customers. Whether that trade-off was worth it is another question. Some monopolies are better than others, as the oil companies say.

No country wants to allow another country, especially one that is not very friendly, to control sections of its infrastructure. Some, like China, manage that themselves, through government ownership. For those who don't, it is a little more complicated because many more companies are involved in all the parts of the service. Besides that, there is intense competition for network equipment provided by several companies that do not provide networks per se. Control of the network components gives an adversary a good start at disrupting network services, through something as simple as shutting off equipment at an appointed time.

Every country has some restrictions on purchasing parts of the telecommunications infrastructure; yet nearly every one of them says they allow other countries to buy into their system. This dichotomy is created by a perception that competition in networking is a good thing that lowers prices, and the belief that having another country own parts of a critical resource is a bad idea.

The federal court broke down the high-level control issues in *Verizon Communications vs. the Federal Communications Commission*, a case involving a ruling by the FCC called the "Open Internet Order." The court said there were four basic types of service affected by the order: backbone networks, broadband providers, edge providers, and end users.

> To pull the whole picture together with a slightly oversimplified example: when an edge provider such as YouTube transmits some sort of content—say, a video of a cat—to an end user, that content is broken down into packets of information, which are carried by the edge provider's local access provider to the backbone network, which transmits these packets to the end user's local access provider, which, in turn, transmits the information to the end user, who then views and hopefully enjoys the cat.[9]

RAND was pointing out the reluctance of government to want to control the Internet, even in time of war, and the District Court of the District of Columbia pointed to some of the complications that go with that. Parts of the infrastructure are regulated, and other parts are not. The Internet is not one thing. It is a series of owners with different roles to play. In the world, China is the only country that tries to control all of the different services that make up the Internet, using a combination of government ownership and restrictive policies. The Russians are trying to get theirs under control, but still have a long way to go. In a free economy, only a monopoly can do the same.

Russia and China have multiple reasons for doing this, but what separates them from the rest of us is the realization that the Internet is not benign, and takes substantial resources to control. In spite of saying it was a CIA project, Putin may well have believed the Internet was controlled by the United States, because the foundations for the Internet rose first in the United States, which started and managed most of the governance functions, even though it was not obligated to do so. In 2014, the United States considered giving up its interest in setting policy for the global Internet through ICANN. It still has made no decision on whether to allow the governance of the Internet to be privatized. There has been a proposal to move governance to the United Nations, but that has not been adopted either. The United States, contrary to popular belief, has no statutory authority over ICANN or the domain name system it governs.[10] It also has no responsibility to support ICANN, even though it historically did.

The Internet, in its current state, is not controllable. Countries want to control those parts that are inside their borders, leading to balkanization, but no country has said it wants to control the whole thing. What countries should do is think about containing the Internet without really trying to control it.

Every country should separate critical functions from the Internet, putting resources into identifying where and how that separation needs to be done, and then doing it. The U.S. Defense Department once ran a couple of exercises to see what not having an Internet connection would do to military functions. They found it impossible to live without. But they made an effort to reduce exposure wherever possible, and made tangible progress at the time. DoD has now probably forgotten they ever did that.

There are three aspects to Internet separation. First, government agencies have to know what they have in their networks (i.e., what network components each has, and how they all interconnect). That requires mapping those components and sharing maps with other organizations they intersect with, something we found to be done only reluctantly by most government agencies and the owners of private networks connecting to them.

Then the agencies have to know what has to be separated from the Internet and aggressively ensure that it is. In 1995, we mapped networks in Huntsville, Alabama, where most of the production for missile defense takes place. Notably, there were over 200 previously unknown connections to the Internet off networks that were not supposed to be connected to it. It took months to seal those off. We found circuits that hadn't been used in over a year, which were terminated. We found contractors connected to government systems despite no longer having contracts with the agencies they were connected to. We found some connections that were terminated, and the government classified the details of those end points. It cost nearly a million dollars and almost nine months of work, but it did give the owners of circuits and computers a good understanding of what they were paying for, what they were responsible for, and how networks can grow almost without conscious activity by anyone.

Third, agencies have to know what functions must be carried out on the Internet. Things like buying supplies, contracting with vendors, communicating with research and university facilities, and the like are all tasks that we would find both expensive and time consuming to do any other way. In the DoD's exercises to see what connectivity they could live without, e-mail with external businesses and educational facilities was the first thing to be identified as a critical need.

Next, we must isolate those tasks that must be completed on the Internet from things that don't have to be. That is done to reduce the ability to use those functions as paths back into the internal government networks. Sometimes, it means air-gapping those connections—that is, putting physical space between the networks (although some hackers have already worked out ways to circumvent air gaps). Defensively, it means using guards, computers made to filter traffic on defined sets of rules, more rigid than firewalls. But the isolation mechanisms have to be controlled, mandated, and maintained.

Last, we need to do more testing of network security in critical functions that we cannot do without. We were very surprised at the amount of testing that was done in Nuclear Command and Control, but probably should not have been. The worst thing that could happen is an unauthorized launch of a nuclear missile, so that testing is justified. The goal of any testing for cyber networks should be to reduce the amount of time it takes to find a successful hack of the network. It now takes days to find one, sometimes longer. In that time, hackers can install other software that will allow them to monitor and modify components all over the network. It takes too long and there is not aggressive follow-up for incidents.

In our early days of doing intrusion detection, we found a hacker getting through firewalls without much trouble. He was getting inside the network and installing software that would send data back out at regular intervals. He was smart enough to only take what was added since his first time in the network. We wanted to know how he was getting the code past the firewall, which was supposed to deny any code from passing though. The Army Research Lab was doing that analysis for us.

It turns out he was not doing it all at once. He was sending what are called code fragments through, one at a time. Our firewalls weren't set up to reject fragments, though we modified the rule sets when that was discovered. It took him days to use a hijacked administrator account to reassemble the code and put it in places where he thought it would be most useful. As an administrator, he could have stolen anything he wanted and hidden his tracks, but he found it more useful to plant codes that would do it for him. He didn't need exposure to the network to do that, and never came back.

In cases like this, we should look for damage that has already been done. One of the things we learned in defending networks was the long-term persistence of attacks. There is an inclination to think of Advanced Persistent Threat as the use of RATs and social engineering to continue getting into systems over a period of time. This is different. When a business or government says, "We have no evidence that any sensitive information was taken" after that hacker got into our systems, they aren't telling the whole story. Once in, hackers erase evidence of their entry and plant Trojan horses to get back in if they are discovered. A 14-year-old kept an air force unit looking for him for weeks before they finally gave up and asked for help. It took six weeks to clean up the mess. Governments are a good deal more capable than this teenager was.

Following clean-up of incidents like the disastrous deployment of the Obamacare website for health insurance or the Chinese hacking the Office of Personnel Management for security clearance information, and at least one of their contractors, there has to be a significant effort to find out what has been done to the system, so hackers can't just walk back in a day after it has been "fixed." Government and industry officials are frequently too quick to declare a system safe, when they should know better.

The civil sector used to do more collaborative defense of its networks. Sometimes it was possible to cooperate with industry sectors independent of governments. Banks and credit card companies probably do more of this kind of collaboration than anyone else, because they are the primary targets of computer crime, but even they don't do enough.

As just one example, there have been several debates about whether Internet banking security is good enough for end users. The answer to this question is not as clear with energy and electricity sectors because end users find it harder to personalize the risk of attacks on those two sectors, even though they have done without electricity and know what it is like. It doesn't last or have the same impact that a credit card theft or bank withdrawal does.

The banks and credit card companies argue that users don't lose money when they experience a loss. That is absorbed by insurance companies and the federal government tax system. But to a user, that argument sounds ridiculous. Hackers are less inclined to go after banks, and prefer to go to end users for access to those accounts.[11] Banks can protect themselves pretty well, but users can't. They have only two choices: take the risk or opt out and drive to the bank. This is where the industry argument—that users don't lose any money—works because users don't have a monetary risk. If we extrapolate that argument, we come to a point where computer crime does no harm to anyone, since insurance companies and governments indemnify the users. That defies logic.

At some point, crime becomes a national security interest, just like theft of intellectual property and damage to critical infrastructures. We are at a point where the intelligence community sees cyber crime as a national security threat, but users and institutions haven't demanded more action. Hackers who profit from this type of crime want to keep the damage within acceptable levels to avoid having governments take more action against them. Yet nobody defines for us what acceptable losses are. They must be acceptable, when nobody says they aren't.

The Russians and Chinese have an advantage over those parts of the world that do not control their Internet. As long as the threshold for war is maintained, as it is in crime, there is no incentive to move toward better defense of networks and deterrence of network attacks. This is not a viable strategy for our national security any more than it is for computer crime.

In past wars, there was time to increase security of national resources during a build-up to actual fighting. Remember the months it took for General Schwarzkopf to put together his forces for the run into Iraq. He had time to wait. Cyberwar does not allow that. The infrastructure targets will have been selected, attacked, and compromised years before an actual attack takes place. The attacker will cripple the industries, people, and critical networks. The attackers also know what the leadership will do. They could tell us they were going to attack, as the Russians did in Crimea, and there would be little we could do about it.

It takes years to develop the human resources used to wage this kind of war. The top companies and government agencies are all competing for that 1 percent of the best who really can do innovative work in defending or attacking networks. Where such candidates exist, they are employed. They aren't waiting for someone to announce a war to come to work.

The Press and Media

In its 1996 study, RAND mentioned two things to the Secretary of Defense, the second one being control of media. In most of the free world, there is only one small area where the media are controlled in any sense of the word. They are licensed, technically, to broadcast on a particular frequency, in a specific geographical area, and licensing is accompanied by content controls. There are some restrictions for obscenity and child pornography, common in most countries. Beyond that, there are few boundaries, and few stations ever lose their licenses.

There is the small area of emergency broadcasting that has equal concern in a few countries. Media are required to carry emergency broadcasts to warn consumers in life-threatening situations. They stop broadcasting regular content to accommodate these messages. Beyond these minor controls, however, the industries are largely self-regulated.

The press heaps criticism on anyone who suggests that journalists, or the media that publish what they write, should be controlled in any way. They claim to be fiercely independent. It is also true that media are somewhat responsive to requests from the leaders or intelligence functions of a government, usually in delaying publication of certain stories for a time, if not forever.

There are abuses on both sides of this delicate balance, but Bill Keller, executive editor of the *New York Times*, summed the issues up best: "A free press in a democracy can be messy. But the alternative is to give the government a veto over what its citizens are allowed to know. Anyone who has worked in countries where the news diet is controlled by the government can sympathize with Thomas Jefferson's oft-quoted remark that he would rather have newspapers without government than government without newspapers."[12]

As a practical matter, no politician will take on the press, but the press has no bounds in what it reports, except the ones it sets. This was particularly true in the book in which Keller wrote the words quoted above. *Open Secrets* is about the work that was done before Bradley (a.k.a. Chelsea) Manning's stolen documents were posted on WikiLeaks. We will probably never know what was not published in those releases, but those things that might have resulted in someone's death were edited or excluded. Still, mistakes were made; it is anything but a science.

In spite of Keller's observation, governments already have a veto on what news the public is allowed to know. Every country has a system of designating certain types of things as national security information. The public has no right to know such information. Most governments use the classification markings *top secret, secret* and *confidential*. What they don't do very well is keep that information out of the hands of press people. We might well wonder what good a system of designating information as protectable is if that information can be published by the press whenever it comes to them.

There are countless documents, still with their top-secret markings, that have been moved to the Internet by people other than Edward Snowden. Each publication does "exceptionally grave damage" to the national security; yet no country has done well at preventing these disclosures, or proposing a solution to the problem.

The five-eyes countries, except for the United States, have an Official Secrets Act, as do a number of other countries. This type of legislation makes it an offense to disclose information that is provided to a newspaper or the press—*if* it falls under the act. The catch is, these acts apply to information that is protected under the laws of that country, not the laws of *any* country. The United States would benefit from a discussion of why other countries, which have far less trouble with reporters publishing national security information, enacted their Official Secrets Acts. When countries share secrets, though a series of bilateral agreements, they need to have an assurance of reciprocity. Most of the WikiLeaks cables, and all of the information Snowden gave up, should have been stopped by agreement between countries that were involved.

There has to be a limit to what the press is allowed to report. We have top-secret, sensitive, compartmented information being published on the Internet, especially in the Snowden releases. We have to balance that exceptionally grave level of damage with what the public is allowed to know. It would be hard for the press to say the public has a right to know what is in these types of documents. It doesn't, and the press should not be allowed to publish this information.

The Chinese are implementing new controls over messaging services, which are inherently difficult to control. The official spokesperson was positively bubbly over the benefits to the average citizen. According to her, we will feel better about sharing information when we use the services supplied by the state. It will be harder for criminals and terrorists to use messages to communicate, she said. No explanation of this particular benefit, or how it might work, was forthcoming. No real reasons were given. Suffice it to say, the controls are a good idea and we will be "in harmony," she said, if we use them.[13]

It is hard to imagine any educated person following along with this kind of blind-faith argument. Maybe it is cultural, but it seems to be something more. In this scenario, the government knows what is good for you. It is almost solicitous. Scary. This is what Big Brother, who now has the technology, has always wanted to be—the benevolent protector of the people.

Maybe this is 1984, the George Orwell version of the future. Every connected government monitors its own population for the reasons this spokeswoman gave: We have to protect the people from criminals and terrorists. We can say the Chinese and Russians do it better than anyone, but we have to wonder if the will of the people is with them.

15

THE NEW WORLD WAR

The Russians found it difficult to fight in the Ukraine, exposing their ambition and lack of patience. They used troops and heavy weapons to shore up their losses in the east and put the entire country at risk. Europe was slow to react, and the U.S. allies did little to help. Perhaps that was the will of the people playing out. Europe and the United States wouldn't step in, and Russia got the benefit of every ugly moment that had casualties mounting. Late in the endgame, the Russians withdrew "most of their forces" and made peace with the Ukraine. In October, they withdrew again. It is almost surreal. Can they really be that good at fighting this kind of war? The proof is coming soon.

Governments make war. Countries don't ask their own people if it was a good idea or something they wanted. Governments make their moves under the guise of national security, where information is separated from the public, confined to policymakers and strategists, as well as those who influence others. Controlling information is essential to operational success. As long as the control of information stays inside their own countries, governments have latitude on how they use the information they acquire.

The United States and its allies are exceptionally good at collecting information about the intentions and plans of other countries or hostile groups. Snowden's disclosures showed how good they really are. They find and share information among themselves, even when they have disagreements about aspects of how it was collected. If you look at terrorism as an indicator, these countries have been remarkably good at finding and stopping the kinds of things that happened on September 11. The vast majority of countries (China and Russia among them) wish they could do as well.

As a rule, we favor what it takes to accomplish this task. Our intelligence services have to collect a good amount of data that has nothing to do with terrorism or crime, sort through it, and act on it. We accept that necessity, in exchange for keeping terrorists from making our countries look like Iraq on a bad day. But there is a matter of degree involved in keeping that from happening, and citizens should have a right to say how far they want that to go. Snowden exposed only one side of what the countries of the world are doing. We need to see what that other side really is.

Russian and Chinese citizens have accepted the level of censorship and control their governments have placed on them, and seem to approve of their current situation. They have no wars, and their economic lives are improving. Why wouldn't they be happy with the way things are going?

They should be content with that level of success. What sets us at war is that they aren't. The world outside of Russia, China, Syria, Iran, and North Korea still believes, at least to some extent, that ideas will win out in the end, if presented well and read by everyone. But those countries want to make sure those ideas are not read by their own populations. The war started when Russia and China began to enforce those access controls outside their own borders.

Only a few countries have the technical capability to wage war at this level. The New World War is between those that do. It is a resource-intensive activity that requires both sophisticated information technology and human analysts who know what to look for. We know that kind of capability can be useful, but it has a potential for misuse. What we have not done well is expose that abuse.

The Germans viewed spying on their head of state, and recruiting government employees to supply information, in that light. But it was not the spying that was at issue as much as being caught at it. The Germans were as hypocritical as the French in this regard, and they have now been accused of spying on communications between the Turks and the United States, and on U.S. diplomats inside Germany.[1] Between allies, lots of things are tolerated that would not be acceptable if exposed. The covert nature of the new war expands that exponentially.

We have the mistaken idea that we can generate words and pictures that show the benefits of our system over the authoritarian governments of the world. As we saw in Crimea, people aren't reading or watching those messages. If we can't expose the tactics of our enemies, we are going to wake up one day and ask ourselves how we managed to get our business and government interests tied into knots by governments that really want to hurt us.

We should be concerned about having these countries involved in our internal politics, attempting to influence our own best sources of information. To each free country, that is an essential element of remaining free. We have to expose what is being done, beyond merely reporting on the occasional hack of a politician's laptop. Who are they influencing and how are they doing it? Where is the money coming from and who benefits from it? The press especially has to have a role in combating what has been done to press sources by those hacking newspapers.

What we have in Russia and China goes further than that. Across the range of elements in cyberwar, both of them have been over the line, Russia more so lately. What the Russians did in Crimea was clever, and we have to give them credit for how they went about it. But we needn't have accepted the final result quite as easily as we did. Maybe we haven't. We certainly shouldn't, since both Russia and China seem willing to increase the attack levels through proxies, and these are going to hurt one day.

China's ally, North Korea, certainly did significant damage to the South Korean economy with its attacks, and the Syrians and Iranians are trying their hand at the same kind of operations. Russia and China have friends like nobody else in the world. They are indicators of something more troubling. If we look closely enough at their tactics, both of them have a kind of in-your-face approach to using other countries' information, and a penchant for inflicting damage through proxies. They use allies to launch the kinds of attacks that are the most destructive. They haven't been very successful yet, because their capabilities are immature. Give them time.

The benefits should be obvious. North Korea fires missiles off the coast of South Korea the day before the Pope arrives to give his first speech there. The Chinese say, "Gee, we wish we could stop that kind of behavior," when everyone knows they can. If Iran or North Korea launches an attack on U.S. banks or the electrical infrastructure, the Russians and Chinese can say it wasn't them. That would be accurate, but can we blame them anyway? We could, and we should, but not without evidence.

There is no deterrent value to anything we do if we can't make the link between cause and effect. This is where U.S. and allied intelligence agencies need to be able to connect the support given to proxies with independent efforts developed by rogue states. That requires intelligence about where and how the techniques are migrating from one country to another, as well as the government officials involved in supporting these kinds of transfers. If we really want to know, we can find out.

We should be sharing that information with the business interests under attack. International commercial business is not doing very well, on its own, in coordinating and sharing it. When it comes to knowing who is stealing proprietary information, plans and strategies, we would think the business community would be focused on discovery and remediation. They should want to know who is doing it, and how they can stop them, as a part of their normal due diligence.

What we are finding instead is a head-in-the-sand view of the world that is reflected in some large businesses appointing Chief Information Security Officers for the first time only after they suffer losses. It is almost like they are discovering, after 20 years of thinking about it, how important a matter this subject is. These days, as we can confirm with anyone in Target,

15. The New World War

Home Depot, T.J. Maxx or Marshalls, the publicity from a major successful hack can do terrific damage to a business's reputation and sales. Strangely, almost the same thing happens to the business reputation of a country when it loses data that another country thinks should be protected.

Among the intelligence services of the world, everyone knows when something is lost that should not have been. It doesn't have to be on the front page of *The Guardian* or *Der Spiegel*. What the Russians did with Snowden causes far more damage to the relationships between allied countries than to anything else. So much information has been given out incrementally that even a slow learner can see that it isn't safe to give anything really sensitive to the United States. Other countries now stop sharing quite as much, and are careful about what they say. That erodes the relationships that take years to build back up.

However, we still have an area in which we can all share with less fear of losing out on sources and methods: We need to get more from our collective intelligence agencies about cyber crime and terrorist financing because both have become national security issues. From the Silk Road kind of website to Afghanistan's support for growing poppies, there are intelligence opportunities in identifying and disrupting drug distribution. Besides funding terrorists, it funds most of the gangs in Latin and South America, as well as Europe and East Asia. Listening to some of the stories of people seeking refugee status in the United States has enlightened many to the scope of gang violence in Central America. The influence of crime continues to grow beyond the countries that house the criminals.

The Israelis have tried to expose some terrorist financing in courts where the proceedings are public, but we have to wonder why that is necessary. Israel isn't the only country concerned with this financing, and certainly not the only one that can expose it. Businesses have to cooperate in getting to the banks and other networks that move money to known terror groups.

Most of all, we need to have more media focus on the kinds of cyberwar campaigns used by Russia and China. We know their press is controlled and, try as it might, would have difficulty publishing stories inconsistent with what their leaders want to see. The compensation for that lies in the free press being free.

Manipulation of the press in other countries should be a press priority. Phony stories about news events that reflect a party line, inconsistent with reality, shouldn't be blindly repeated. Publicizing of the abuse of journalists is a start. Exposure of some of the crude lines of attack, like Russia's use of Nazi symbolism, or a claim that dead bodies were placed on a civilian airliner and it was shot down on purpose, has to be done more often. Repeating the story without refuting the claim is not good journalism. That goes for politics in our own countries as much as for international uses. It is partly laziness, and lack of focus, that allows this to happen.

We should know who among us would fight a cyberwar and make sure they can. It won't be the armies of those countries involved. In the United States it is the intelligence services. In Great Britain it is MI6; in Australia, the Secret Intelligence Service. In Russia it is the Federal Security Service (FSB); in China, the Ministry of State Security (MSS). These are the intelligence services, and they fight the new wars. They lead in policy and execution. These are the only agencies that know how to fight clandestine wars, and are equipped to do it. If militaries participate, they do so under the direction of the intelligence services.

We already know that most governments favor the collection of intelligence over the defense of their computer networks. They want the intelligence services making decisions on when and how cyberwar is carried out, so it doesn't interfere with those operations. The way to do that is establish the intelligence services as policy leaders, which almost every country in the world has done. It should be little wonder that we fight more of our wars in the dark, considering who is fighting them.

Cyberwar has a long way to go, but if you had asked someone in 1995 how big it was going to be, they would have said "not very." Wars have been fought for centuries, but the way they have been fought in the past 20 years will change warfare forever. We are just beginning to see the possibilities.

Chapter Notes

Introduction

1. CNN *News Day*, Cable News Network, 28 August 2014.
2. Mao Tse-tung, "Be Concerned with the Well-Being of the Masses, Pay Attention to Methods of Work" (January 27, 1934), China Books & Periodicals, Inc., San Francisco, p. 226.

Chapter 1

1. Paul Sonne, Philip Shishkin, and Anton Troianovsk, "Shift Toward Moscow Jolts Crimean Economy," *Wall Street Journal*, 17 March 2014.
2. IANS, "Hackers Attack Russian Media Outlets," *Business Standard*, 18 April 2014.
3. Guy Chazan and Courtney Weaver, "Russia's Return," *Financial Times Weekend*, 22/23 March 2014.
4. David E. Sanger and Steven Erlanger, "Suspicion Falls on Russia as 'Snake' Cyberattacks Target Ukraine's Government," *New York Times*, 8 March 2014.
5. Lester Grau and Timothy Thomas, "Russian Lessons Learned in the Battles for Grozny," *Marine Corps Gazette*, April 2000.
6. Jonathan Marcus, "Ukraine: The Military Balance of Power," BBC News, 3 March 2014.
7. Stephen Fidler, "NATO Releases More Photos It Says Show Russian Troop Buildup Near Ukraine," *Wall Street Journal*, 11 April 2014.
8. Security Service of the Ukraine, "A Military Spy Detained" (in English), 14 March 2014, http://www.sbu.gov.ua/sbu/control/en/publish/article?art_id=122723&cat_id=35317.
9. Sam Jones, "Evidence Points to Insurgent Puppetmaster," *Financial Times*, 9 August 2014.
10. Ewen MacAskill, "Does U.S. Evidence Prove Russian Special Forces Are in Eastern Ukraine?" *The Guardian*, 22 April 2014.
11. Jack Farchy, "Putin Faces Mother of All Battles at Home," *Financial Times*, 30–31 August 2014.
12. Arwa Damon, "Ukrainian Official's Body Found in River," CNN, 22 April 2014.
13. Eric Brown, "White House: Eastern Ukrainian Protesters May Have Been Paid by Russia," *International Business Times*, 7 April 2014.
14. Lucy Crossley, "The 'Aggrieved Housewife,' the 'Soldiers Mother' and the 'Kiev Resident': Did Russian Television Use Actress to Portray FIVE Different Women as It Reported Normal Ukrainians Backed Kremlin?" *London Daily Mail* (online), 5 March 2014.
15. David M. Herszenhorn and Andrew Roth, "In East Ukraine, Protesters Seek Russian Troops," *New York Times*, 7 April 2014.
16. Andrew Kramer and Michael Gordon, "Russia Sent Tanks to Separatists in Ukraine, U.S. Says," *New York Times*, 13 June 2014.
17. ITAR-TASS News Agency, "Court to Hear Arrest of Ukraine's Interior Minister in Absentia Behind Closed Doors," 9 July 2014.
18. See discussion on methods and means of warfare, International Committee of the Red Cross, http://www.icrc.org/eng/war-and-law/conduct-hostilities/methods-means-warfare/overview-methods-and-means-of-warfare.htm.
19. Alan Cullison, "Ukraine's Secret Weapon: Feisty Oligarch Ihor Kolomoisky," *Wall Street Journal*, 27 June 2014.

20. ITAR-TASS News Agency, "Hackers Crack PrivatBank Systems for Sponsoring Military Campaign in Ukraine," 30 June 2014.
21. Kathrin Hille, "Russia Bars Ukraine Banks from Crimea," *Financial Times*, 21 April 2014.
22. Glenn Kates, "Ukraine's Minister of Facebook," Radio Free Europe, 9 July 2014.
23. Peter Finn, "Russia-Georgia War Intensifies," *Washington Post*, 10 August 2008.
24. Alexander Tanas, "Moldova Blocks Russian Plan to Expand Presence in Rebel Enclave," Reuters, 17 November 2012.
25. Christian Oliver, "Transnistria Feels Squeezed by Clash Between East and West," *Financial Times*, 5/6 April 2014.
26. ISH Janes, "Ratification of Russian Military Base Deal Provides Tajikistan with Important Security Guarantees," Janes 360, 1 October 2013, http://www.janes.com/article/27898/ratification-of-russian-military-base-deal-provides-tajikistan-with-important-security-guarantees.
27. LizFuller,"Chechen Leader Gives Exclusive Interview to RFE/RL," Radio Free Europe, 7 March 2005.
28. Philip Taubman, "Soviets List Afghan War Toll: 13,310 Dead, 35,478 Wounded," *New York Times*, 26 May 1988.
29. Peter Baker and Ellen Barry, "U.S. Penalizes Russians for Human Rights Violations," *New York Times*, 12 April 2013.
30. Matthew Karnitschnig, "German Businesses Urge Halt on Sanctions Against Russia," *Wall Street Journal*, 1 May 2014.
31. Courtney Weaver, "Moscow Launches Mac Attack as Beef with West Over Ukraine Intensifies," *Financial Times*, 26–27 July 2014.
32. FoxNews.com, "U.S. Troops Arrive in Poland for Exercises Across Eastern Europe Amid Ukraine Crisis," 23 April 2014.
33. Reuters, "Update 3—Russia Targets Space Station Project in Retaliation for U.S. Sanctions," 13 May 2014.
34. Dermot O'Shea, "Real-World Drive Tests Declare a Verdict on GPS/GLONASS," *Electronic Design*, 31 May 2013.
35. Jay Solomon, "Ties to Russia Arms Supplier Snarl U.S. Sanctions Effort," *Wall Street Journal*, 28 March 2014.
36. Sharon Weinberger, "How to Do Business with a Blacklisted Russian Weapons Company," *Wired*, 28 July 2008.
37. Maria Vassilieva, "Russian Elections: Hunting the 'Carousel' Voters," *BBC Russia*, 5 March 2012.
38. See Department of Defense Directive 5200-1R, Chapter 8, http://www.fas.org/irp/doddir/dod/5200-1r/chapter_8.htm.
39. Peter Bergen, "Who Really Killed Bin Laden?" *CNN World News*, 27 March 2013, http://www.cnn.com/2013/03/26/world/bergen-who-killed-bin-laden/.
40. Associated Press, "Ukraine Clashes Kill 2 as Russia Ramps Up Military Exercises," *CBCNews*, 24 April 2014.
41. Sam Jones, "Photos and Roses for GRU's 'Spetsnaz' Casualties," *Financial Times*, 8 August 2014.
42. Robert Chesney, "Military-Intelligence and the Law of the Title 10/Title 50 Debate," University of Texas School of Law, *Journal of National Security Law and Policy* 5 (2012): 539.
43. BBC News, "Ukraine Crisis: What the 'Russian Soldier' Photos Say," 22 April 2014.
44. CNBC Squawk Box, 11 September 2014.
45. Substantively the same in every major Russian news outlet: *Voice of Russia*, "U.S. Starts World Press Freedom Day with Even More Lies," 6 May 2014.

Chapter 2

1. The Cold War Files, "Leonid M. Kravchuk," Woodrow Wilson International Center for Scholars, Cold War International History Project.
2. John-Thor Dahlburg, "Ukraine Votes to Quit Soviet Union: Independence: More than 90% of Voters Approve Historic Break with Kremlin," *Los Angeles Times*, 3 December 1991, and "News Analysis: Bush's 'Chicken Kiev' Talk an Ill-Fated U.S. Policy," *Los Angeles Times*, 19 December 1991.
3. Elizabeth Shogren, "Tensions Rise Over Fate of Black Sea Fleet," *Los Angeles Times*, 4 April 1992.
4. Doyle McManus, "U.S. to Press Ukrainian on Nuclear Issue," *Los Angles Times*, 6 May 1992.
5. Steven Pifer, "Ukraine's Perpetual East-West Balancing Act," The Brookings Institute, 28 February 2014.
6. The Russian press spells his name Yanukovich in their articles in English.
7. Peter Finn, "Yushchenko Was Poisoned, Doctors Say," *Washington Post*, 12 December 2004.
8. Leslie MacKinnon, "The Mystery of Irwin Cotler's Poisoning in Russia," *CBC*, 1 April 2014.

9. Ian Traynor, "Russia Accused of Unleashing Cyberwar to Disable Estonia," *The Guardian*, 16 May 2007.

10. Press Release, "Ukrtelecom's Crimean Sub-branches Officially Report That Unknown People Have Seized Several Telecommunications Nodes in Crimea," 28 February 2014.

11. David Lee, "Russia and Ukraine in Cyber 'Stand-off,'" *BBC News Technology Report*, 5 March 2014.

12. Information Warfare Monitor and Shadowserver Foundation, "Shadows in the Cloud: Investigating Cyber Espionage 2.0," 6 April 2010.

13. Lo Ping Cheng Ming, "Secrets About CPC Spies—Tens of Thousands of Them Scattered Over 170-Odd Cities Worldwide," No. 231, 1 January 1997, 6–9 ["Journal Discloses 'Secrets' About PRC Spy Network," FBIS-CHI-97-016, 1 January 1997]. See http://www.fas.org/irp/world/china/mss/budget.htm for a range of documents on Chinese collection efforts. See Dan Raviv and Yossi Melman, *Every Spy a Prince* (Boston: Houghton Mifflin, 1990).

14. Christopher Andrew and Vasili Mitrokhin, *The Sword and the Shield* (New York: Basic Books, 1999), 213–267.

15. Cheryl Kim, "Think You Are a Thought Leader? You Are Probably Wrong … but Here Are Three Ways to Become One," *Financial Post*, 7 March 2014.

16. Cellular News, "Russia's Rostelecom Urged to Expand Services into Crimea," 25 March 2014.

17. Mark Clayton, "What Crimea Telecom Link Could Mean for Russia-Ukraine Cyber-Conflict," *Christian Science Monitor*, 7 April 2014.

18. Promotional piece at http://www.epegcable.com/#page-partners.

19. Promotional piece, "Analysis: DREAM Project Expected to Enhance Europe-Asia Connectivity," *Capacity* magazine, 2013, http://www.capacitymagazine.com/Article/3267837/ANALYSIS-DREAM-project-expected-to-enhance-Europe-Asia-connectivity.html.

20. Laura Hedges, "Tele2 Russia to Deploy LTE Network in Crimea," *Capacity* magazine, 2 April 2014.

21. International Telecommunications Union (Geneva), "Percentage of Individuals Using the Internet 2010–2012," June 2013, retrieved 22 June 2013.

22. Peter Pomerantsev, "How Putin Is Reinventing Warfare," *Foreign Policy*, 5 May 2014.

23. Declan Walsh, "Pakistan Suspends License of Leading News Channel," *New York Times*, 7 June 2014.

24. Ellen Barry and Michael Schwirtz, "After Election Putin Faces Challenges to Legitimacy," *New York Times*, 5 March 2012.

25. Mark Clayton, "Ukraine Election Narrowly Avoided 'Wanton Destruction' from Hackers," *Christian Science Monitor*, 17 June 2014.

26. Andrew and Mitrokhin, 243. See also BBC News, "Ukraine-E.U. Trade Deal 'Big Threat' to Russia's Economy," 26 November 2013.

27. BBC News, "Ukraine-E.U. Trade Deal," and BBC News, *Ukraine Crisis Timeline*, http://www.bbc.com/news/world-Europe-25108022, 11 April 2014.

28. The Committee to Protect Journalists has several stories on its website including "Ukrainian Journalists Held by Pro-Russian Separatists," https://cpj.org/2014/07/ukrainian-journalists-held-by-pro-russian-separati.php.

29. RT, unattributed Russian article with video: "Yanukovich Sent Letter to Putin Asking for Russian Military Presence in Ukraine," 3 March 2014.

30. Jon Boyle, "Ukraine Hit by Cyberattacks: Head of Ukraine Security Service," Reuters, 4 March 2014.

31. Conal Urquhart, "Summary of Ukraine Events," *The Guardian*, 1 March 2014.

32. AFT/PTI, "Crimea Hit with Partial Power Outages," *Business Standard*, 24 March 2014.

33. Lauri Lowenthan Marcus, "'Jews Must Register' Flyer in Ukraine an Echo of Babi Yar," *Jewish Press*, 18 April 2014.

34. Steven Erlanger, "Russia Ratchets Up Ukraine's Gas Bills in Shift to an Economic Battlefield," *New York Times*, 11 May 2014.

35. Alastair Beach, "Russian Oligarchs Fear Further Sanctions Over Ukraine," *The Telegraph*, 21 July 2014.

36. Lee S. Wolosky, "How to Sanction Russia," Foreign Affairs, Council on Foreign Relations, 19 March 2014.

37. Devlin Barrett, Christopher Matthews and Andrew Johnson, "BNP Paribas Draws Record Fine for 'Tour de Fraud,'" *Wall Street Journal*, 30 June 2014.

38. Natalie Robehmed, "U.S. Treasury

Sanctions Four Russian Billionaires for Ukraine Involvement," *Forbes*, March 2014.

39. Jon Greenburg, "John Kerry Tells CBS Viewers That Russian Isolation Comes with Big Economic Risks," *Tampa Bay Times*, undated, http://www.politifact.com/truth-o-meter/statements/2014/mar/02/john-kerry/john-kerry-tells-cbs-viewers-russian-isolation-com/.

40. William Mauldin and Paul Sonne, "U.S. Escalates Sanctions Against Russia Over Ukraine Crisis," *Wall Street Journal*, 16 July 2014.

41. Marinelog, "U.S. Sanctions Putin's Shipbuilder and Its Bank," accessed 30 July 2014, http://www.marinelog.com/index.php?option=com_k2&view=item&id=7072:us-sanctions-putins-shipbuilder-and-its-bank&Itemid=231.

42. Office of Foreign Asset Control, Sectoral Sanctions Identifications List, 29 July 2014.

43. Office of Foreign Asset Control; Wolosky.

44. Michael Weiss, Cable News Network, a Turner Broadcasting System of Time Warner, 18 July 2014.

45. James Marson, "Russian Muscle Turns Tide, Leaving Ukraine Few Options," *Wall Street Journal*, 3 September 2014.

Chapter 3

1. Carl von Clausewitz, *On War* (Penguin Books, original 1832), 101.

2. Ibid., 123.

3. Stockholm International Peace Research Institute publishes an annual report on armed conflict and related issues. See http://www.sipri.org/yearbook/2013/01.

4. James Blight, Janet Lang, et al., *Becoming Enemies* (Lanham, MD: Rowman & Littlefield, 2012), 28–29.

5. Public Broadcasting Service, *The Iranian Hostage Crisis*, undated, http://www.pbs.org/wgbh/americanexperience/features/general-article/carter-hostage-crisis/.

6. Raymond H. Anderson, "Ayatollah Ruhollah Khomeini, 89, the Unwavering Iranian Spiritual Leader," *New York Times*, 4 June 1989.

7. Roger Hardy, "The Iran-Iraq War: 25 Years On," *BBC News*, 22 September 2005.

8. The History of Iran, *The Iran-Iraq War 1980–1988*, Iran Chamber Society, http://www.iranchamber.com/history/iran_iraq_war/iran_iraq_war1.php.

9. Simon Tisdall, "Kurdistan Faces Long, Fraught Road to Sustainable Independence," *The Guardian*, 3 July 2014.

10. Efraim Karsh, *The Iran-Iraq War: 1980–1988* (Oxford: Osprey, 2002), 12–16.

11. Shane Harris and Mathew Aid, "Exclusive: CIA Files Prove America Helped Saddam as He Gassed Iran," *Foreign Policy*, 26 August 2013.

12. Joseph P. Fried, "Gerald Bull, 62, Shot in Belgium; Scientist Who Violated Arms Law," *New York Times*, 25 March 1990.

13. The Central Intelligence Agency, *Project Babylon: The Iraqi Supergun*, CIA Electronic Reading Room, November 1991.

14. Wafic al Samarrai, *Frontline: The Gulf War (Oral History)*, aired 9 January 1996, copyright 1995–2014, WGBH Educational Foundation.

15. Ibid.

16. Norman Schwarzkopf, *It Doesn't Take a Hero* (New York: Bantam Books, 1992), 301.

17. Michael R. Gordon and General Bernard E. Trainor, *Cobra II* (New York: Pantheon Books, 2006), 164–165.

18. Jeffrey Richelson (ed.), *Iraq and Weapons of Mass Destruction*, National Security Archives, updated 11 February 2004, http://www2.gwu.edu/~nsarchiv/NSAEBB/NSAEBB80/.

19. "The 9/11 Commission Report: Identifying and Preventing Terrorist Financing," 23 August 2004, p. 47.

20. Regan Doherty and Amena Bakr, "Exclusive: Secret Turkish Nerve Center Leads Aid to Syria Rebels," Reuters, 27 July 2012.

21. Mark Hosenball, "Exclusive: Obama Authorizes Secret U.S. Support for Syrian Rebels," Reuters, 1 August 2012.

22. Elizabeth O'Bagy, "The Free Syrian Army," Institute for the Study of War, 2013.

23. Patrick Cockburn, "Who Are ISIS? The Rise of the Islamic State in Iraq and the Levant," *The Independent* and *The Times of India*, 19 June 2014.

24. See Deutsche Welle, "Thousands Flee Mosul in Iraq as ISIL Takes Control," 11 June 2014.

25. David Gauthier-Villars, "France Calls for Action to Cut Off ISIS Money Supply," *Wall Street Journal*, 22 August 2014.

26. Jane Arraf, "The Resurrection of Ahmad Chalabi," *Foreign Policy*, 10 July 2014.

Chapter 4

1. Various news reports throughout the day from Cable News Network, Turner Broadcasting System of Time Warner, 18 July 2014.
2. Masha Alekhina, "Russian Media Is Covering Up Putin's Complicity in the MH17 Tragedy," *The Guardian*, 19 July 2014.
3. Al Jazeera America is a Delaware Corporation broadcasting on several U.S. channels. Various coverage from 18–20 July 2014.
4. Tabatha Kinder, "Israel Uses Image of London Missile Strike to Appeal for Gaza Support," *International Business Times*, 21 July 2014.
5. Joshua Mitnick, "Israelis Frustrated with Outcome of Gaza Conflict," *Wall Street Journal*, updated 29 August 2014.
6. James R. Clapper, "The Worldwide Threat Assessment of the U.S. Intelligence Community," Senate Select Committee on Intelligence, 29 January 2014.
7. See Intelligence and Security Committee Annual Report to Parliament, July 2013, pages 8–10.
8. Jane Perlez, "China Shows Interest in Afghan Security, Fearing Taliban Would Help Separatists," *New York Times*, 8 June 2012.
9. Alex Hern, "North Korean 'Cyberwarfare' Said to Have Cost South Korea L500M," *The Guardian*, 16 October 2013.
10. Spencer Ante, "IBM, Lenovo Tackle Security Worries on Server Deal," *Wall Street Journal*, 25 June 2014.
11. Frederick W. Whatley, *Reagan, National Security, and the First Amendment: Plugging Leaks by Shutting Off the Main* (Washington, DC: Cato Institute, 1984).
12. Te-Ping Chen, "Snowden Alleges U.S. Hacking in China," *Wall Street Journal*, updated 23 June 2013.
13. Matthew Goldstein, "Silk Road Case Began with Hunt for a John Doe," *New York Times*, 21 March 2014.
14. Clive Thompson, "The Darkest Place on the Internet Isn't Just for Criminals," *Wired*, 18 October 2013.
15. John Daly, "Iran Tears Up Azadegan Contract with China," Oilprice.com, 4 May 2014, http://oilprice.com/Energy/Energy-General/Iran-Tears-Up-Azadegan-Contact-With-China.html.
16. Alexei Anishchuk, "As Putin Looks East, China and Russia Sign $400-Billion Gas Deal," Reuters, 21 May 2014.
17. Alexandra Thomsen, "The Disease Daily," *Healthmap*, July 2014.
18. BBC News, Stuxnet "Hit" Iran Nuclear Plans, 22 November 2010, http://www.bbc.com/news/technology-11809827.
19. Richard Weitz, "Superpower Symbiosis: The Russia-China Axis," *World Affairs Journal*, November–December 2012.
20. Jane Perlez, "China and Russia, in a Display of Unity, Hold Naval Exercises," *New York Times*, 10 July 2013.
21. Weitz.
22. William Burr, ed., "The Sino-Soviet Border Conflict 1969: U.S. Reactions and Diplomatic Maneuvers," published 12 June 2001 by the National Security Archive, George Washington University.
23. WGBH Educational Affiliation, *Nixon's China Game, American Experience*, Public Broadcasting, 1999.
24. Weitz, 47.
25. Romina Boccia, Allison Acosta Fraser, and Emily Goff, "Federal Spending by the Numbers, 2013: Government Spending Trends in Graphics, Tables and Key Points," Heritage Foundation, 20 August 2013, http://www.heritage.org/research/reports/2013/08/federal-spending-by-the-numbers-2013.
26. Ben Blanchard and John Ruwitch, "China Hikes Defense Budget to Spend More on Internal Security," Reuters, 5 March 2013.
27. Agence France-Presse, "Iran Plans 127% Defense Budget Increase," *Defense News*, 2 February 2012, http://www.defensenews.com/article/20120202/DEFREG04/302020003/.
28. Amos Harel and Gili Cohen, "Israeli Military Seeks Massive Top-up of 2014 Defense Budget," *Haaretz*, 6 December 2013, http://www.haaretz.com/news/diplomacy-defense/.premium-1.555217.
29. United Press International, "Russia Plans Increases in Spending on Defense, Nuclear Weapons," 8 October 2013, http://www.upi.com/Top_News/World-News/2013/10/08/Russia-plans-increases-in-spending-on-defense-nuclear-weapons/UPI-36261381251991/.
30. Thom Shanker, "U.S. Arms Sales Make Up Most of Global Market," *New York Times*, 26 August 2012, http://www.nytimes.com/2012/08/27/world/middleeast/us-foreign-arms-sales-reach-66-3-billion-in-2011.html.

31. Arthur M. Schlesinger Jr., "The Measure of Diplomacy," *Foreign Affairs*, July/August 1994, http://www.foreignaffairs.com/articles/50118/arthur-m-schlesinger-jr/the-measure-of-diplomacy.

Chapter 5

1. See the Internet Society Studies, Global Internet User Survey 2012, http://www.internetsociety.org/surveyexplorer/key_findings.
2. Vodaphone, Law Enforcement Disclosure Report, 2014, http://www.vodafone.com/content/sustainabilityreport/2014/index/operating_responsibly/privacy_and_security/law_enforcement.html#eocp.
3. Barton Gellman, "Obama's Restrictions on NSA Surveillance Rely on Narrow Definition of 'Spying,'" *Washington Post*, 17 January 2014.
4. Charles Doyle and Jennifer Elsea, "The Posse Comitatus Act and Related Matters: The Use of the Military to Execute Civilian Law," Congressional Research Office, 16 August 2012.
5. Brian A. Jackson, *Considering the Creation of a Domestic Intelligence Agency in the United States* (Santa Monica, CA: RAND Corporation, 2009), 9.
6. Vodaphone.
7. Agence France-Press, "Rolling Stones Told Not to Play Honky Tonk Women at Shanghai Gig," *The Guardian*, 13 March 2014.
8. BBC News, "China Detains Xinjiang Man for 'Online Rumours,'" 11 August 2014.
9. Gary King, Jennifer Pan, and Margaret Roberts, "How Censorship in China Allows Government Criticism but Silences Collective Expression," *American Political Science Review* 107, no. 2 (May): 1–18.
10. Andrei Soldatov, "Vladimir Putin's Cyber Warriors," *Foreign Affairs*, 9 December 2011.
11. Timothy Heritage, "Putin Dissolves State News Agency, Tightens Grip on Russia Media," Reuters, 9 December 2013.
12. Anton Troianovski, "Russia Ramps Up Information War in Europe," *Wall Street Journal*, 21 August 2014.
13. RIA Novosti, "Rossiya Segodnya's First Ever Arabic Newswire Launching Monday, September 22," 18 September 2014.
14. Annie Mortensen, "A Journalist Goes Missing Nearly Every Day in Ukraine," *Independent Voices*, 27 May 2014.
15. Dan Friedman, "Twitter Stepping Up Suspensions of ISIS-Affiliated Accounts: Experts," *New York Daily News*, 17 August 2014.
16. Jacob Silverman, "Loose Tweets Sink Ships," *Politico*, 28 August 2014.
17. Evgeny Morozov and Joanne J. Myers, "The Net Delusion: The Dark Side of Internet Freedom," Carnegie Council, 25 January 2011.
18. John D. Sutter, "When the Internet Actually Helps Dictators," Cable News Network, 22 February 2011.
19. Mohammed El-Nawawy and Sahar Khamis, "Political Activism 2.0: Comparing the Role of Social Media in Egypt's 'Facebook Revolution' and Iran's 'Twitter Uprising,'" *CyberOrient* 6, issue 1 (2012), http://www.cyberorient.net/article.do?articleId=7439.
20. Guy Chazan and Courtney Weaver, "Russia's Return," *Financial Times*, 22/23 March 2014.
21. James Bamford, in a 2014 *Wired* article cited in a later entry, says Snowden blames this outage on the United States.
22. Ellen Nakashima, "Report: Web Monitoring Devices Made by U.S. Firm Blue Coat Detected in Iran, Sudan," *Washington Post*, 8 July 2013.
23. Steve Stecklow, "Dubai Firm Fined $2.8 Million for Shipping Blue Coat Monitoring Gear to Syria," Reuters, 25 April 2013. For a more technical analysis, see Morgan Marquis-Boire, Jakub Dalek, et al., "Planet Blue Coat: Mapping Global Censorship and Surveillance Tools," *The Citizen Lab*, 15 January 2013.
24. Marquis-Boire, Dalek, et al.
25. RIA Novosti, "Two-Thirds of Russians Are Internet Users," 21 May 2013. (Curiously, the headline for the article says 2/3 of the users, but the data shown does not support that.)
26. Andrei Soldatov and Irina Borogan, "The Kremlin's Internet Surveillance Plan Goes Live Today," *Wired*, 1 November 2012.
27. Tomasz Bujlow, Valentín Carela-Español, and Pere Barlet-Ros, "Comparison of Deep Packet Inspection (DPI) Tools for Traffic Classification," Technical Report, Universitat Politecnica de Catalunya, 30 June 2013.
28. Greg Walton, "China's Golden Shield," International Center for Human Rights and Development, 2001.
29. Jonathan Zittrain and Benjamin Edel-

man, "Empirical Analysis of Internet Filtering in China," Harvard Law School, March 2003.

30. Openet Initiative, "China's Green Dam: The Implications of Government Control Encroaching on the Home PC," https://opennet.net/chinas-green-dam-the-implications-government-control-encroaching-home-pc.

31. See Bruce Schneier at Congress for Privacy and Security, http://www.youtube.com/watch?v=Skr-jIqISO0.

32. Amnesty International Report, "Iran: New Report Finds Surge in Repression of Dissent," 28 February 2012.

33. Associated Press, "Hernandez Challenges Evidence," ESPN video, http://espn.go.com/boston/nfl/story/_/id/11091660/aaron-hernandez-challenges-evidence-odin-lloyd-murder-case.

34. Seth Hardy, "IExplore RAT," *Citizen Lab Technical Brief*, August 2012, https://citizenlab.org/wp-content/uploads/2012/09/IEXPL0RE_RAT.pdf.

35. See Anonymous Symantec Employee, "Emerging Threat: Dragonfly/Energetic Bear—APT Group," Symantec Corp, 30 June 2014, http://www.symantec.com/connect/blogs/emerging-threat-dragonfly-energetic-bear-apt-group.

36. Dominique Haughton et al., "A Review of Software Packages for Data Mining," *The American Statistician* 57, no. 4 (November 2003).

37. King, Pan, and Roberts, 1.

38. Lauren Weber, "At Work: Remote Data Wipes of Workers' Personal Devices Are Rising," *Wall Street Journal*, 9 September 2014.

39. Roxana Geambasu, Tadayoshi Kohno, Amit A. Levy, and Henry M. Levy, "Vanish: Increasing Data Privacy with Self-Destructing Data," University of Washington, http://css.csail.mit.edu/6.858/2014/readings/vanish.pdf.

40. Curtis Cartier, "Narus, Boeing-Owned Company, Is Helping Egyptian Government's Web Crackdown in Cairo," *Seattle News Weekly*, 28 January 2011.

41. Michael J. Lee et al., "Network Discovery with Multi-Intelligence Sources," *Lincoln Laboratory Journal* 20, no. 1 (2013).

42. Shadowserver Foundation Report AS40989, "RBN as RBusiness Network," 6 January 2008.

43. Brian Krebs, "Shadowy Russian Firm Seen as Conduit for Cybercrime," *Washington Post*, 13 October 2013.

44. Brian Krebs, "Russian Business Network, Down but Not Out," *Washington Post*, 7 November 2007.

45. Bruce Sterling, "Russian Business Network Sets Up Shop in China," *Wired*, 9 November 2013.

46. Geoff McDonald et al., "Stuxnet 0.5: The Missing Link," Symantec Security Response, 23 January 2014.

47. David Sanger, "Obama Order Sped Up Wave of Cyberattacks Against Iran," *New York Times*, 1 June 2012.

48. Doug Bernard, "Russia Tightens Grip on the Internet," Voice of America, 27 April 2014.

49. Allissa de Carbonnel and Gerry Shih, "Russia Asks Twitter to Block a Dozen Accounts," Reuters, 24 June 2014.

50. Kathrin Hille, "Russian Bill Raises Web Freedom Fears," *Financial Times*, 5–6 July 2014.

51. Nicole Perlroth, "Hackers in China Attacked the Times for the Last 4 Months," *New York Times*, 30 January 2013.

52. Nicole Perlroth, "*Washington Post* Joins List of News Media Hacked by the Chinese," *Washington Post*, 1 February 2013.

53. Scott Harold and Alireza Nader, "China and Iran—Economic, Political, and Military Relations," RAND Corporation, 2012.

54. Nicole Perlroth, "In Cyberattack on Saudi Firm, U.S. Sees Iran Firing Back," *New York Times*, 23 October 2012.

55. Stephen Ward, "An Iranian Threat Inside Social Media," iSight Partners, 28 May 2014.

56. Ellen Nakashima, "Pentagon Is Debating Cyber-Attacks," *Washington Post*, 6 November 2010.

Chapter 6

1. See Anonymous Symantec Employee, "Emerging Threat: Dragonfly/Energetic Bear—APT Group," Symantec Corp, 30 June 2014, http://www.symantec.com/connect/blogs/emerging-threat-dragonfly-energetic-bear-apt-group.

2. Michael Vatis, "Cyber Attacks: Protecting America's Security Against Digital Threats," ESDP Discussion Paper ESDP-2002-04, John F. Kennedy School of Government, Harvard University, June 2002.

3. Nicole Perlroth and Quentin Hardy, "Bank Hacking Was the Work of Iranians, Officials Say," *New York Times*, 8 January 2013.

Chapter 7

1. Jacob Bunge, "U.S. Arrests Second Chinese Citizen in Seed-Theft Case," *Wall Street Journal*, 2 July 2014.
2. Ellen Nakashima, "U.S. Said to Be Target of Massive Cyber-Espionage Campaign," *Washington Post*, 10 February 2013.
3. Fareed Zakaria, "Russian and Chinese Assertiveness Poses New Foreign Policy Challenges," HBO History Makers Series, Council on Foreign Relations, 21 May 2014.
4. Adam Rawnsley, "Espionage? Moi?" *Foreign Policy*, 1 July 2013.
5. Jack Devine, *Good Hunting* (New York: Sarah Crichton Books/Farrar, Straus, and Giroux, 2014), 197.
6. "APT1," 23–24.
7. Nathan Thornburgh, "Inside the Chinese Hack Attack," *Time*, 25 August 2005.
8. Adam Elkus, "Moonlight Maze," in *A Fierce Domain: Conflict in Cyberspace, 1986–2012*, edited by Jason Healy (Vienna, VA: Cyber Conflict Studies Association, 2013).
9. CBS News, *60 Minutes*, "Huawei Probed for Security, Espionage Risk," 7 October 2012.
10. Ernesto Londofio, "As U.S. Withdraws from Afghanistan, Poppy Trade It Spent Billions Fighting Still Flourishes," *Washington Post*, 3 November 2013; Master Sgt. Jeff Szczechowski, USAF, "Security Forces Help Keep Convoys Safe in Afghanistan," American Forces Press Service, 26 May 2004.
11. RT (RT is a Russian-based television network, previously known as Russia Today), "U.S. Seen More Risky for Industrial Espionage: German Study," 6 August 2013.
12. Richards J. Heuer Jr., *Psychology of Intelligence Analysis*, Central Intelligence Agency, Chapter 4, 16 March 2007.

Chapter 8

1. David E. Sanger and Steven Erlanger, "Suspicion Falls on Russia as 'Snake' Cyberattacks Target Ukraine's Government," *New York Times*, 8 March 2014.
2. BAE Applied Intelligence, "Snake Campaign and Cyber Espionage Toolkit," 2014, p. 5.
3. National Security Agency, Pearl Harbor Review, JN-25, http://www.nsa.gov/about/cryptologic_heritage/center_crypt_history/pearl_harbor_review/jn25.shtml.
4. Jess Bravin, "Echoes from a Past Leak Probe," *Wall Street Journal*, 7 August 2013.
5. Greg Miller, "Video Shows Brazen Outdoor Meeting of al-Qaeda Fighters," *Washington Post*, 15 April 2014.
6. See YouTube stats at http://www.youtube.com/yt/press/.
7. Sony Pictures, 2013. (The White House helped in the making of this movie.)
8. Al Jazeera and the Associated Press, "U.S. Drone Strike Kills New Zealander, Australian in Yemen," 16 April 2014.
9. Reuters and Al Jazeera, "Yemen Drones Strikes, Ambushes Kill 10," 3 March 2014.
10. Mohammed Jamjoom, "Source: 'Massive' Attack on Targets al Qaeda in Yemen," CNN, 20 April 2014.
11. Milton Hoenig, "Hezbollah and the Use of Drones as a Weapon of Terrorism," Federation of American Scientists, 5 June 2014.

Chapter 9

1. Stefan Wagstul, "Steinmeier Feels Pressure as Outburst Goes Viral Online," *Financial Times*, 23 May 2014.
2. "World Development Indicators," The World Bank Group, 2014.
3. Ibid.
4. Pernilla Holmes, "The Art of Photojournalism," HowtoSpendIt (website of the *Financial Times*), 7 April 2014, http://howtospendit.ft.com/art/51443-the-art-of-photojournalism.
5. Central Intelligence Agency, "Afghanistan: Communications," *CIA Factbook*, 2012.
6. "Literacy by Country," World by Map, http://world.bymap.org/LiteracyRates.html.
7. Everette E. Dennis, Justin D. Martin, and Robb Wood, "Media Use in the Middle-East: An Eight-Nation Survey," Northwestern University in Qatar, April 2013.
8. Matt Bradley, "Egyptian TV Swayed Public Against Morsi, in Favor of Sisi," *Wall Street Journal*, 28 May 2014.
9. Ann T. Greeley, "Psychology, Technology, and the Art of Expert Witness Persuasion in the Internet Age," *Dispute Resolution Insights* (Summer 2011): 72.
10. Dan Murphy, "U.S. Nabs Alleged Russian Hacker—and Kremlin Cries Foul," *Christian Science Monitor*, 9 July 2014.
11. ITAR/TASS, "Russian Lawmaker Val-

ery Seleznyov Confirms Detention of His Son by U.S. Secret Service," 8 July 2014.

Chapter 10

1. John Prados and Svetlana Savranskaya, eds., *Volume II: Afghanistan: Lessons from the Last War*, The National Security Archive (2001).
2. Associated Press, "Chechen Rebels Hurting for Money," September 2004, http://www.foxnews.com/story/0,2933,132617,00.html.
3. Suadad Al-Salhy, "Iraqi Shi'ites Flock to Assad's Side as Sectarian Split Widens," Reuters, 19 June 2013, reuters.com/article/2013/06/19/us-iraq-syria-militants; Farnaz Fassihi, "Iran Pays Afghans to Fight for Assad," *Wall Street Journal*, 15 May 2014.
4. Drew Hinshaw, Adama Diakite, and Stacy Meichtry, "France Faces Revival of Mali Militants," *Wall Street Journal*, 19 December 2013, http://online.wsj.com/news/article_email/SB10001424052702303330204579246103662992322-lMyQjAxMTAzMDIwMDEyNDAyWj.
5. Libya Television, "Kidal: Lastest Tuareg Mercenaries (MIA, MNLA) Preparing for Ultimate Battle," Al Jazeera English, 13 May 2013.
6. Pew Forum on Religious and Public Life, December 2012, http://www.pewforum.org/2012/12/18/global-religious-landscape-exec/.
7. Caleb MacLeod, "No Guns, Just Knives: Chilling Details of 'China's 9/11,'" *USA Today*, 30 March 2014.
8. Jeremy Page, "Suspect Shown on Chinese TV Confessing to Ax Attack," *Wall Street Journal*, 22 June 2014.
9. Jonathan Kaiman, "Six Injured in Knife Attack at Chinese Train Station," *The Guardian*, 6 May 2014.
10. Michael Martina, "Chinese State Media Says Five Suicide Bombers Carried Out Xinjian Attack," Reuters, 23 May 2014.
11. Jeremy Page and Ned Levin, "Web Preaches Jihad to China's Muslim Uighurs," *Wall Street Journal*, 24 June 2014.
12. Gaye Christoffersen, "Constituting the Uyghur in U.S.-China Relations: The Geopolitics of Identity Formation in the War on Terrorism," Center for Contemporary Conflict, Naval Postgraduate School, 2 September 2002.
13. Matthew Levitt and Michael Jacobson, "The Money Trail: Finding, Following and Freezing Terrorist Financing," The Washington Institute, November 2008.
14. BBC News, "China Urges Pakistan to Expel Uighur Islamic Militants," 31 May 2012.
15. Alison Frankel, "Israel's Conflicted Role in Bank of China Terror Finance Case," Reuters, U.S. Edition, 11 August 2014.
16. See Osen LLC, Counter-Terrorism, *Linde v. Arab Bank*, http://www.osenlaw.com/case/arab-bank-case.
17. Joe Palazzolo, "Terror Victims to Press Claims Against Arab Bank," *Wall Street Journal*, 8 August 2014.
18. See http://www.osenlaw.com/case/arab-bank-case.
19. There are several organizations that use the name Saudi Committee, followed by suffixes of clarification.
20. U.S. Department of Treasury, "U.S. Designates Five Charities Funding Hamas and Six Senior Hamas Leaders as Terrorist Entities," 22 August 2003.
21. Jean MacKenzie, "Who Is Funding the Afghan Taliban? You Don't Want to Know," Reuters, 13 August 2009.
22. Quinton Summerville, "Pakistan Helping Afghan Taliban—NATO," BBC News, 1 February 2012.
23. James R. Clapper, "Worldwide Threat Assessment to the House Permanent Select Committee on Intelligence," 4 February 2014.
24. This is an alternate spelling for Al Qaeda, which the State Department uses.
25. U.S. Treasury Department Public Release, "Treasury Targets Key Al-Qa'ida Funding and Support Network Using Iran as a Critical Transit Point," 28 July 2011.
26. "The 9/11 Commission Report: Identifying and Preventing Terrorist Financing," 23 August 2004.
27. Ron Synovitz, "Tracking the Terrorist Money Trail," Radio Free Europe/Radio Liberty, November 2008, http://www.rferl.org/content/Interview_Tracking_The_Terrorist_Money_Trail/1349657.html.
28. Juan C. Zarate and Thomas Sanderson, "How the Terrorists Got Rich," *New York Times*, 28 June 2014.
29. Rachel Ehrenfeld, "Where Hamas Gets Its Money," *Forbes*, 16 January 2009.
30. Vladimir Lenin, *War and Revolution*, Pravda, 23 April 1917, as published in *Lenin Collected Works* Vol. 24 (Moscow: Progress Publishers, 1964), 398–421, and on the Lenin Internet Archive, 2005.

31. BBC News, "'Anti-Semitic' French Envoy Under Fire," 20 December 2001, http://news.bbc.co.uk/2/hi/1721172.stm.
32. Jeremy M. Sharp, "U.S. Foreign Aid to Israel," Congressional Research Service, 11 April 2013, http://www.fas.org/sgp/crs/mideast/RL33222.pdf.
33. The *Forward* and Josh Nathan-Kazis, "Uncovering the U.S. Jewish Charity Industries," *Harretz*, 26 March 2014.
34. Pew Forum on Religious and Public Life.
35. Elizabeth Shell and Matt Stiles, "Where Does U.S. Military Aid Go?" McNeil/Lehrer Productions, 30 August 2012, http://www.pbs.org/newshour/spc/multimedia/military-spending/.
36. Michael Maretinez, Talal Abu Rahma, and Kareem Khadder, "Israel Fires on 29 'Terror Sites' After Rockets from Gaza Hit Populated Areas," CNN, 12 March 2014.
37. Wolf Blitzer, CNN News, 7 July 2014.
38. Amos Harel and Gili Cohen, "Reports: Israeli Planes Attack Hezbollah Targets on Lebanon-Syria Border," *Haaretz*, 25 February 2014.
39. Jay Solomon, "U.S. Blacklists Companies That Supplied Hezbollah," *Wall Street Journal*, 11 July 2014.
40. CNN News, 7 July 2014.
41. Saad Abedine and Michael Schwartz, "Israel Intercepts Ship with Weapons Headed to Gaza," CNN, 6 March 2014.
42. Drew Desilver, "Worlds Muslim Population More Widespread Than You Might Think," Pew Research Center, 7 June 2013.
43. Jessica Morris, "Iran-Saudi Relations: A New Cold War Heating Up?" CNBC News, 28 January 2014.
44. Steven Erlanger, "Iran and 6 Powers Agree on Terms for Nuclear Talks," *New York Times*, 20 February 2014.
45. R. Jeffrey Smith and Joby Warrick, "Pakistani Scientist Khan Describes Iranian Efforts to Buy Nuclear Bombs," *Washington Post*, March 14, 2010, and Christopher Clary, *The A. Q. KHAN Network: Causes and Implications*, Naval Post Graduate School, December 2005.

Chapter 11

1. CBS News, "Computer Glitch Halts Metro-North Trains for Nearly 2 Hours," 23 January 2014.
2. Eric Torbenson, "ATC Problems on East Coast: Not Much Here," *Dallas Morning News*, 19 November 2009.
3. David Sanger, David Barboza, and Nicole Perlroth, "Chinese Army Unit Is Seen as Tied to Hacking Against U.S.," *New York Times*, 18 February 2013.
4. Kevin Ashton, "That 'Internet of Things' Thing," *RFID*, 22 June 2009.
5. Florida Center for Instructional Technology, *Internet Basics*, University of South Florida, updated 2009.
6. Robert Marquand and Ben Arnoldy, "China's Hacking Skills in Spotlight," *Christian Science Monitor*, 16 September 2007.
7. John P. Mello Jr., "Pentagon: Yep, We Got Hacked," *Tech News World*, 26 August 2010.
8. Nick Hopkins, "Hackers Have Breached Top Secret MoD Systems, Cybersecurity Chief Admits," *The Guardian*, 3 May 2012.

Chapter 12

1. BBC News, "Gazprom Hikes Ukraine Gas Price by a Third," BBC News, 1 April 2014.
2. Tom Parfitt, "Russian Media Ignore WikiLeaks Revelations," *The Guardian*, 2 December 2010.
3. Robert M. Gates, *Duty* (New York: Alfred A. Knopf, 2014), 412.
4. Scott P. Boylan, "Organized Crime and Corruption in Russia: Implications for U.S. and International Law," *Fordham International Law Journal* 19 (1999).
5. Holly Ellyatt, "Is Russia Too Corrupt for International Business?" CNBC, 11 June 2013.
6. Steven Ike, "KGB Influence 'Soars Under Putin,'" BBC News, 13 December 2006.
7. Keith Moor, "Russian Crime Gang's $570M Cyber Theft Bid as Nation Loses $4.6B a Year to Computer Crime," *The Herald Sun*, 23 July 2014.
8. Charles Clover, "Who Runs Russia?" *Financial Times*, 16 December 2011.
9. See Immigration and Protection Tribunal case number [2012] NZIPT 800151 at http://www.refworld.org/pdfid/50321fd92.pdf.
10. Charles Clover and Tom Mitchell, "Senior Chinese General Falls Victim to Xi's Corruption Crackdown," *Financial Times*, 30 June 2014.
11. Jeremy Page, Brian Spegele, and

James Areddy, "China Pus Ex-Security Chief Zhou YongKang Under Investigation," *Wall Street Journal*, 29 July 2014.
   12. RIA Novosti, "Corruption Up 450% in a Year in Russian Military—Prosecutors," 7 November 2013.
   13. Isabel Gorst, "Russian Military Budget Sapped by Corruption," *Financial Times*, 24 May 2011.
   14. U.S. Army, "Mobilization," undated, page 21, http://www.history.army.mil/documents/mobpam.htm.
   15. Office of the Secretary of Defense, "DoD Website Defense Manpower Report," https://www.dmdc.osd.mil/appj/dwp/reports.do?category=reports&subCat=milActDutReg.
   16. Moshe Schwartz, "Twenty-Five Years of Acquisition Reform: Where Do We Go from Here?" Congressional Research Service, 29 October 2013, page 1.
   17. Phillip Swartz, "Lawmakers Force Pentagon to Buy Tanks, Keep Ships and Planes It Doesn't Need," *Washington Times*, 9 May 2013.
   18. Associated Press, "Congress Forcing Military to Keep Unwanted Assets, Programs Despite Spending Cuts, Report Says," 23 April 2013.
   19. Jonathan Skillings, "Airborne Laser Hits the Off Switch," CNET, 27 February 2012.
   20. Robert F. Hale, "Promoting Efficiency in the Department of Defense: Keep Trying, Be Realistic," Center for Strategic and Budgetary Assessments, January 2002, p. 7.
   21. The Stockholm International Peace Research Institute, "The SIPRI Top 100 Arms-Producing and Military Services Companies in the World Excluding China, 2011," 2013, http://www.sipri.org/research/armaments/production/Top100.
   22. Thom Shanker, "U.S. Arms Sales Make Up Most of Global Market," *New York Times*, 26 August 2012.
   23. Lockheed Martin Annual Report, 2012, p. 12.
   24. Raytheon Annual Report, 2012, p. 15.
   25. Doug Cameron, "Lockheed Martin Posts Higher Profit, Sees Relief on Pension Front," *Wall Street Journal*, 22 April 2014.
   26. United Press International Business News, "DSCA Outlines Foreign Military Sales Program," 20 September 2013, http://www.upi.com/Business_News/Security-Industry/2013/09/20/DSCA-outlines-foreign-military-sales-program/UPI-415013797114 18/-ixzz2uGUs4X00.
   27. Deloitte Development LLC, "2013 Global Aerospace and Defense Industry Outlook," https://www.deloitte.com/assets/Dcom-Iceland/Local%20Assets/Documents/2013GlobalADIndustryOutlook.pdf.
   28. Lawrence R. Jacobs and Benjamin I. Page, "Who Influences U.S. Foreign Policy?" *American Political Science Review* 99, no. 1 (February 2005).
   29. Alan Greenspan, *The Map and the Territory* (New York: Penguin Press, 2013), 266.
   30. Gara Afonso, João Santos, and James Traina, "Large and Complex Banks," *Economic Policy Review* 20, no. 2, Federal Reserve of New York (March 2014): 2.
   31. John Sweeney, Jens Holsoe and Ed Vulliamy, "NATO Bombed Chinese Deliberately," *The Observer*, 16 October 1999.
   32. Doug Cameron, "Cut Weapons Now, But Then What?" *Wall Street Journal*, 3 July 2014.
   33. Amy Belasco, "The Cost of Iraq, Afghanistan, and Other Global War on Terror Operations Since 9/11," Congressional Budget Office, 29 March 2011, p. 1.
   34. Jack Devine, *Good Hunting* (New York: Sarah Crichton Books/Farrar, Straus, and Giroux, 2014).
   35. Zachary Keck, "China Is a Leading Proliferator of Small Arms," *The Diplomat*, October 2013, http://thediplomat.com/2013/10/china-is-a-leading-proliferator-of-small-arms/.
   36. Amnesty International, "The 'Big Six' Arms Exporters," June 2012, http://www.amnesty.org/en/news/big-six-arms-exporters-2012-06-11.
   37. The Stockholm International Peace Research Institute, "The SIPRI Yearbook," 2013, p. 10.
   38. C.J. Chivers and Eric Schmitt, "Arms Shipments Seen from Sudan to Syria Rebels," *New York Times*, 12 August 2013.
   39. Ilya Gridneff, "China Sells South Sudan Arms as Its Government Talks Peace," *Bloomburg News*, 9 July 2014.
   40. Small Arms Survey as quoted in "Sudan's Military Industry Expanding: Small Arms Survey," Radio Dabanga (Sudan), 6 July 2014.
   41. Jim Finkle and Andrea Shalal-Esa, "Lockheed Network Suffers Major Disruption—Sources," Reuters U.S., 26 May 2011.

42. Brian Krebs, "Hackers Plundered Israeli Defense Firms That Built 'Iron Dome' Missile Defense System," KrebsonSecurity, 14 July 2014, http://krebsonsecurity.com/2014/07/hackers-plundered-israeli-defense-firms-that-built-iron-dome-missile-defense-system/.
43. GAO, "Federally Funded Research Centers: Agency Reviews of Employee Compensation and Center Performance," 10 September 2014.
44. James Lewis and Stewart Baker, "The Economic Impact of Cyber Crime and Cyber Espionage," Center for International and Strategic Studies, July 2013, pp. 3 and 5.
45. Glenn Greenwald and Ewen MacAskill, "Obama Orders U.S. to Draw Up Overseas Target List for Cyber-attacks," *The Guardian*, 7 June 2013.
46. Ibid.
47. Lisa Fleisher, "U.K. Says Laws Permit Mass Spying on Residents Online," *Wall Street Journal*, 17 June 2014.

Chapter 13

1. Janet Reitman, "The Rise and Fall of Jeremy Hammond: Enemy of the State," *Rolling Stone*, 7 December 2012, 4 and 6.
2. U.S. Department of Justice, Press Release, "Manhattan U.S. Attorney Announces Guilty Plea of Jeremy Hammond for Hacking into the Stratfor Website," 28 May 2013.
3. Christian Fotinger and Wolfgang Ziegler, "Understanding a Hacker's Mind—A Psychological Insight into the Hijacking of Identities," Danube-University and RSA Security, Inc.
4. Nart Villeneuve, Ned Moran and Thoufique Haq, "Operation Molerats: Middle East Cyber Attacks Using Poison Ivy," *FireEye*, 23 August 2013.
5. *Guardian* Staff Report, "Edward Snowden and the NSA Files—Timeline," *The Guardian*, 23 June 2013.
6. James Bamford, "The Most Wanted Man in the World," *Wired*, 13 June 2014.
7. Adam Sneed, "Rogers: Russia May Be Behind Snowden Leaks," *Politico*, 14 January 2014.
8. National Security Agency, Cryptologic Almanac 50th Anniversary Series, Betrayers of Trust, undated, http://www.nsa.gov/public_info/_files/crypto_almanac_50th/Betrayers_of_the_Trust.pdf.
9. Michael Walsh, "Unmasked," Smithsonian Institution, May 2009.
10. John Prados, "The Navy's Biggest Betrayal," *Naval History Magazine* 24, no. 3 (June 2010).
11. CIA Archives, "The People of the CIA ... Ames Mole Hunt Team," 12 March 2009, updated 30 April 2013.
12. Reuters Staff Report, "Edward Snowden Tells Der Spiegel NSA Is 'in Bed with the Germans,'" *The Guardian*, 7 July 2013; Ewen MacAskill, Julian Borger, Nick Hopkins, et al., "GCHQ Taps Fibre-optic Cables for Secret Access to World's Communications," *The Guardian*, 21 June 2013.
13. Jill Reilly, "The Guardian Allowed GCHQ Spies into Its Offices to Oversee Destruction of Leaked NSA Data After 'Senior Government Figure Backed by Prime Minister' Threatened Legal Action," *The Daily Mail*, 20 August 2013.

Chapter 14

1. Roger Molander, Andrew Riddle, and Peter Wilson, "Strategic Information Warfare," RAND, 1996.
2. Robert Gates, Council on Foreign Relations.
3. Pew Research Global Attitude Project, "Global Opinion of Russia Mixed," 3 September 2013, http://www.pewglobal.org/2013/09/03/global-opinion-of-russia-mixed/.
4. Pew Research Center, "Far More Continue to View Russia as 'a Serious Problem' Than as an 'Adversary,'" 28 July 2014, http://www.people-press.org/2014/07/28/far-more-continue-to-view-russia-as-a-serious-problem-than-as-an-adversary/.
5. John Clifton, "150 Million Adults Worldwide Would Migrate to the U.S.," Gallup, 2012, http://www.gallup.com/poll/153992/150-million-adults-worldwide-migrate.aspx.
6. Christopher J. Bright, "Cold War Air Defense Relied on Widespread Dispersal of Nuclear Weapons, Documents Show," The National Security Archive, George Washington University, posted 16 November 2010.
7. See "Nuclear Weapons Effects Technology Militarily Critical Technologies List, Part II: Weapons of Mass Destruction Technologies," at the Federation of American Scientists website, http://fas.org/nuke/intro/nuke/effects.htm, accessed 2 August 2014.

8. Richard Waters, "Attacks Lead to New Line for Telcos in U.S.," *Financial Times*, 12 October 2001.
9. United States Court of Appeals, District of Columbia Circuit, *Verizon v. Federal Communications Commission & Independent Telephone & Telecommunications Alliance, ET AL*, 14 January 2014.
10. Lennard G. Kruger, "Internet Governance and the Domain Name System: Issues for Congress," Congressional Research Service, 10 June 2014.
11. Lisa Bachelor, "On-line Banking Fraud Losses Rise 14%," *The Guardian*, 10 March 2010.
12. Alexander Star, ed., *Open Secrets* (New York: Grove Press, 2011), 17.
13. *China Take*, Bon Television Network, aired 11 August 2014.

Chapter 15

1. BBC News, "Germany Accused of Spying on Kerry and Clinton," 16 August 2014; Associated Press, "Turkey Protests Germany's Spying on U.S.-Turkey communications," *Epoch Times*, 18 August 2014.

Bibliography

Abedine, Saad, and Michael Schwartz. "Israel Intercepts Ship with Weapons Headed to Gaza." CNN, 6 March 2014.
Afonso, Gara, João Santos, and James Traina. "Large and Complex Banks," *Economic Policy Review* 20, no. 2, Federal Reserve of New York (March 2014).
AFT/PTI. "Crimea Hit with Partial Power Outages," *Business Standard*, 24 March 2014.
Agence France-Presse. "Iran Plans 127% Defense Budget Increase." *Defense News*, 2 February 2012.
———. "Rolling Stones Told Not to Play Honky Tonk Women at Shanghai Gig." *The Guardian*, 13 March 2014.
Alekhina, Masha. "Russian Media Is Covering Up Putin's Complicity in the MH17 Tragedy." *The Guardian*, 19 July 2014.
Al Jazeera and the Associated Press. "U.S. Drone Strike Kills New Zealander, Australian in Yemen." 16 April 2014.
Al-Salhy, Suadad. "Iraqi Shi'ites Flock to Assad's Side as Sectarian Split Widens." Reuters, 19 June 2013, reuters.com/article/2013/06/19/us-iraq-syria-militants.
Amnesty International. "The 'Big Six' Arms Exporters." June 2012. http://www.amnesty.org/en/news/big-six-arms-exporters-2012-06-11.
———. "Iran: New Report Finds Surge in Repression of Dissent." 28 February 2012.
Anderson, Raymond H. "Ayatollah Ruhollah Khomeini, 89, the Unwavering Iranian Spiritual Leader," *New York Times*, 4 June 1989.
Andrew, Christopher, and Vasili Mitrokhin. *The Sword and the Shield*. New York: Basic Books, 1999.
Anishchuk, Alexei. "As Putin Looks East, China and Russia Sign $400-Billion Gas Deal." Reuters, 21 May 2014.
Ante, Spencer. "IBM, Lenovo Tackle Security Worries on Server Deal." *Wall Street Journal*, 25 June 2014.
Ashton, Kevin. "That 'Internet of Things' Thing." *RFID*, 22 June 2009.
Associated Press. "Chechen Rebels Hurting for Money." September 2004.
———. "Congress Forcing Military to Keep Unwanted Assets, Programs Despite Spending Cuts, Report Says." 23 April 2013.
———. "Hernandez Challenges Evidence." ESPN video. http://espn.go.com/boston/nfl/story/_/id/11091660/aaron-hernandez-challenges-evidence-odin-lloyd-murder-case.
———. "Turkey Protests Germany's Spying on U.S.-Turkey Communications." *Epoch Times*, 18 August 2014.
———. "Ukraine Clashes Kill 2 as Russia Ramps Up Military Exercises." *CBCNews*, 24 April 2014.
BAE Applied Intelligence. "Snake Campaign and Cyber Espionage Toolkit." 2014.

Baker, Peter, and Ellen Barry. "U.S. Penalizes Russians for Human Rights Violations." *New York Times*, 12 April 2013.
Bamford, James. "The Most Wanted Man in the World." *Wired*, 13 June 2014.
Barrett, Devlin, Christopher Matthews and Andrew Johnson. "BNP Paribas Draws Record Fine for 'Tour de Fraud.'" *Wall Street Journal*, 30 June 2014.
Barry, Ellen, and Michael Schwirtz. "After Election Putin Faces Challenges to Legitimacy." *New York Times*, 5 March 2012.
Bachelor, Lisa. "On-line Banking Fraud Losses Rise 14%," *The Guardian*, 10 March 2010.
BBC News. "Anti-Semitic' French Envoy Under Fire." 20 December 2001. http://news.bbc.co.uk/2/hi/1721172.stm.
———. "China Detains Xinjiang Man for 'Online Rumours.'" 11 August 2014.
———. "China Urges Pakistan to Expel Uighur Islamic Militants." 31 May 2012.
———. "Gazprom Hikes Ukraine Gas Price by a Third." 1 April 2014.
———. "Germany Accused of Spying on Kerry and Clinton." 16 August 2014.
———. "Ukraine Crisis: What the 'Russian Soldier' Photos Say." 22 April 2014.
———. *Ukraine Crisis Timeline*. http://www.bbc.com/news/world-europe-25108022, 11 April 2014.
———. "Ukraine-E.U. Trade Deal 'Big Threat' to Russia's Economy." 26 November 2013.
Beach, Alastair. "Russian Oligarchs Fear Further Sanctions Over Ukraine." *The Telegraph*, 21 July 2014.
Belasco, Amy. "The Cost of Iraq, Afghanistan, and Other Global War on Terror Operations Since 9/11." Congressional Budget Office, 29 March 2011.
Bergen, Peter. "Who Really Killed Bin Laden?" *CNN World News*, 27 March 2013. http://www.cnn.com/2013/03/26/world/bergen-who-killed-bin-laden/.
Bernard, Doug. "Russia Tightens Grip on the Internet." Voice of America, 27 April 2014.
Blanchard, Ben, and John Ruwitch. "China Hikes Defense Budget to Spend More on Internal Security." Reuters, 5 March 2013.
Blight, James G., Janet M. Lang, et al. *Becoming Enemies: U.S.-Iran Relations and the Iran-Iraq War, 1979–1988*. Lanham, MD: Rowman & Littlefield, 2014.
Blitzer, Wolf. CNN News, 7 July 2014.
Boccia, Romina, Allison Acosta Fraser, and Emily Goff. "Federal Spending by the Numbers, 2013: Government Spending Trends in Graphics, Tables and Key Points." Heritage Foundation, 20 August 2013.
Boylan, Scott P. "Organized Crime and Corruption in Russia: Implications for U.S. and International Law." *Fordham International Law Journal* 19 (1999).
Boyle, Jon. "Ukraine Hit by Cyberattacks: Head of Ukraine Security Service." Reuters, 4 March 2014.
Bradley, Matt. "Egyptian TV Swayed Public Against Morsi, in Favor of Sisi." *Wall Street Journal*, 28 May 2014.
Bravin, Jess. "Echoes from a Past Leak Probe." *Wall Street Journal*, 7 August 2013.
Bright, Christopher J. "Cold War Air Defense Relied on Widespread Dispersal of Nuclear Weapons, Documents Show," The National Security Archive—George Washington University, posted 16 November 2010.
Brown, Eric. "White House: Eastern Ukrainian Protesters May Have Been Paid by Russia." *International Business Times*, 7 April 2014.
Bujlow, Tomasz, Valentín Carela-Español, and Pere Barlet-Ros. "Comparison of Deep Packet Inspection (DPI) Tools for Traffic Classification." Technical Report, Universitat Politecnica de Catalunya, 30 June 2013.
Bunge, Jacob. "U.S. Arrests Second Chinese Citizen in Seed-Theft Case." *Wall Street Journal*, 2 July 2014.
Burns, Christopher. "Espionage Takes Toll on Air Show; U.S. Firms on Guard." Associated Press, 14 June 1993.

Bibliography

Burr, William, ed. "The Sino-Soviet Border Conflict 1969: U.S. Reactions and Diplomatic Maneuvers." National Security Archive, George Washington University, 12 June 2001.
Callam, Andrew. "Drone Wars: Armed, Unmanned Aerial Vehicles." *International Affairs Review* XVIII, no. 3 (Winter 2010).
Cameron, Doug. "Cut Weapons Now, but Then What?" *Wall Street Journal*, 3 July 2014.
_____. "Lockheed Martin Posts Higher Profit, Sees Relief on Pension Front." *Wall Street Journal*, 22 April 2014.
Cartier, Curtis. "Narus, Boeing-Owned Company, Is Helping Egyptian Government's Web Crackdown in Cairo." *Seattle News Weekly*, 28 January 2011.
CBS News. "Computer Glitch Halts Metro-North Trains for Nearly 2 Hours." 23 January 2014.
Cellular News. "Russia's Rostelecom Urged to Expand Services into Crimea." 25 March 2014.
Central Intelligence Agency. "Afghanistan: Communications." *CIA Factbook*. 2012.
Chazan, Guy, and Courtney Weaver. "Russia's Return." *Financial Times Weekend*, 22/23 March 2014.
Chen, Te-Ping. "Snowden Alleges U.S. Hacking in China." *Wall Street Journal*, updated 23 June 2013.
Chesney, Robert. "Military-Intelligence and the Law of the Title 10/Title 50 Debate." University of Texas School of Law, *Journal of National Security Law and Policy* 5 (2012): 539.
Chivers, C.J., and Eric Schmitt. "Arms Shipments Seen from Sudan to Syria Rebels." *New York Times*, 12 August 2013.
Christoffersen, Gaye. "Constituting the Uyghur in U.S.-China Relations: The Geopolitics of Identity Formation in the War on Terrorism." Center for Contemporary Conflict, Naval Postgraduate School, 2 September 2002.
CIA Archives. "The People of the CIA ... Ames Mole Hunt Team." 12 March 2009, updated 30 April 2013.
Clapper, James R. "The Worldwide Threat Assessment of the U.S. Intelligence Community." Senate Select Committee on Intelligence, 29 January 2014.
_____. "Worldwide Threat Assessment to the House Permanent Select Committee on Intelligence." 4 February 2014.
Clary, Christopher. *The A. Q. KHAN Network: Causes and Implications*. Naval Post Graduate School, December 2005.
Clausewitz, Carl von. *On War*. London, UK: Penguin, 1832.
Clayton, Mark. "Ukraine Election Narrowly Avoided 'Wanton Destruction' from Hackers." *Christian Science Monitor*, 17 June 2014.
_____. "What Crimea Telecom Link Could Mean for Russia-Ukraine Cyber-Conflict." *Christian Science Monitor*, 7 April 2014.
Clifton, John. "150 Million Adults Worldwide Would Migrate to the U.S." Gallup, 2012. http://www.gallup.com/poll/153992/150-million-adults-worldwide-migrate.aspx.
Clover, Charles. "Who Runs Russia?" *Financial Times*, 16 December 2011.
_____, and Tom Mitchell. "Senior Chinese General Falls Victim to Xi's Corruption Crackdown." *Financial Times*, 30 June 2014.
CNN *New Day*, Cable News Network, 28 August 2014.
Cockburn, Patrick. "Who Are ISIS? The Rise of the Islamic State in Iraq and the Levant," *The Independent* and *The Times of India*, 19 June 2014.
Crossley, Lucy. "The 'Aggrieved Housewife,' the 'Soldiers Mother' and the 'Kiev Resident': Did Russian Television Use Actress to Portray FIVE Different Women as It Reported Normal Ukrainians Backed Kremlin?" *London Daily Mail* (online), 5 March 2014.
The Cold War Files. "Leonid M. Kravchuk." Woodrow Wilson International Center for Scholars, Cold War International History Project.

Bibliography

Cullison, Alan. "Ukraine's Secret Weapon: Feisty Oligarch Ihor Kolomoisky." *Wall Street Journal*, 27 June 2014.
Dahlburg, John-Thor. "News Analysis: Bush's 'Chicken Kiev' Talk an Ill-Fated U.S. Policy." *Los Angeles Times*, 19 December 1991.
_____. "Ukraine Votes to Quit Soviet Union: Independence: More than 90% of Voters Approve Historic Break with Kremlin." *Los Angeles Times*, 3 December 1991.
Daily Mail Reporter. "Israel Overtakes America as the World's Largest Jewish Population Center for the First Time." *London Daily Mail*, 29 March 2013.
Daly, John. "Iran Tears Up Azadegan Contract with China." Oilprice.com, 4 May 2014. http://oilprice.com/Energy/Energy-General/Iran-Tears-Up-Azadegan-Contact-With-China.html.
Damon, Arwa. "Ukrainian Official's Body Found in River." CNN, 22 April 2014.
de Carbonnel, Allissa, and Gerry Shih. "Russia Asks Twitter to Block a Dozen Accounts." Reuters, 24 June 2014.
Deloitte Development LLC. "2013 Global Aerospace and Defense Industry Outlook." https://www.deloitte.com/assets/Dcom-Iceland/Local%20Assets/Documents/2013GlobalADIndustryOutlook.pdf.
Dennis, Everette E., Justin D. Martin, and Robb Wood. "Media Use in the Middle-East: An Eight-Nation Survey." Northwestern University in Qatar, April 2013.
Department of Defense Directive 5200–1R, Chapter 8. http://www.fas.org/irp/doddir/dod/5200-1r/chapter_8.htm.
Desilver, Drew. "Worlds Muslim Population More Widespread Than You Might Think." Pew Research Center, 7 June 2013.
Deutsche Welle. "Thousands flee Mosul in Iraq as ISIL takes control," 6 November 2014. http://www.dw.de/thousands-flee-mosul-in-iraq-as-isil-takes-control/a-17698724.
Devine, Jack. *Good Hunting*. New York: Sarah Crichton Books/Farrar, Straus, and Giroux, 2014.
Doherty, Regan, and Amena Bakr. "Exclusive: Secret Turkish Nerve Center Leads Aid to Syria Rebels" *Reuters*, 27 July 2012.
Doyle, Charles, and Jennifer Elsea. "The Posse Comitatus Act and Related Matters: The Use of the Military to Execute Civilian Law." Congressional Research Office, 16 August 2012.
Ehrenfeld, Rachel. "Where Hamas Gets Its Money." *Forbes*, 16 January 2009.
Elkus, Adam. "Moonlight Maze." *A Fierce Domain: Conflict in Cyberspace, 1986–2012*, edited by Jason Healy. Vienna, VA: Cyber Conflict Studies Association, 2013.
Ellyatt, Holly. "Is Russia Too Corrupt for International Business?" CNBC, 11 June 2013.
El-Nawawy, Mohammed, and Sahar Khamis. "Political Activism 2.0: Comparing the Role of Social Media in Egypt's 'Facebook Revolution' and Iran's 'Twitter Uprising.'" *CyberOrient* 6, issue 1 (2012). http://www.cyberorient.net/article.do?articleId=7439.
Erlanger, Steven. "Iran and 6 Powers Agree on Terms for Nuclear Talks." *New York Times*, 20 February 2014.
_____. "Russia Ratchets Up Ukraine's Gas Bills in Shift to an Economic Battlefield." *New York Times*, 11 May 2014.
Farchy, Jack. "Putin Faces Mother of All Battles at Home." *Financial Times*, 30–31 August 2014.
Fassihi, Farnaz. "Iran Pays Afghans to Fight for Assad." *Wall Street Journal*, 15 May 2014.
Fidler, Stephen. "NATO Releases More Photos It Says Show Russian Troop Buildup Near Ukraine." *Wall Street Journal*, 11 April 2014.
Finkle, Jim, and Andrea Shalal-Esa. "Lockheed Network Suffers Major Disruption—Sources." Reuters U.S., 26 May 2011.
Finn, Peter. "Russia-Georgia War Intensifies." *Washington Post*, 10 August 2008.
_____. "Yushchenko Was Poisoned, Doctors Say." *Washington Post*, 12 December 2004.

Fleisher, Lisa. "U.K. Says Laws Permit Mass Spying on Residents Online." *Wall Street Journal*, 17 June 2014.
Florida Center for Instructional Technology. *Internet Basics*. University of South Florida, updated 2009.
The *Forward* and Josh Nathan-Kazis. "Uncovering the U.S. Jewish Charity Industries." *Harretz*, 26 March 2014.
Fotinger, Christian, and Wolfgang Ziegler. "Understanding a Hacker's Mind—A Psychological Insight into the Hijacking of Identities." Danube-University and RSA Security, Inc.
FoxNews.com. "U.S. Troops Arrive in Poland for Exercises Across Eastern Europe Amid Ukraine Crisis." 23 April 2014.
Frankel, Alison. "Israel's Conflicted Role in Bank of China Terror Finance Case." Reuters, U.S. Edition, 11 August 2014.
Fried, Joesph P. "Gerald Bull, 62, Shot in Belgium; Scientist Who Violated Arms Law." *New York Times*, 25 March 1990.
Friedman, Dan. "Twitter Stepping Up Suspensions of ISIS-Affiliated Accounts: Experts." *New York Daily News*, 17 August 2014.
Fuller, Liz. "Chechen Leader Gives Exclusive Interview to RFE/RL." Radio Free Europe, 7 March 2005.
GAO. "Federally Funded Research Centers: Agency Reviews of Employee Compensation and Center Performance." 10 September 2014.
Gates, Robert M. *Duty*. New York: Alfred A. Knopf, 2014.
Gellman, Barton. "Obama's Restrictions on NSA Surveillance Rely on Narrow Definition of 'Spying.'" *Washington Post*, 17 January 2014.
Goldstein, Matthew. "Silk Road Case Began with Hunt for a John Doe." *New York Times*, 21 March 2014.
Gordon, Michael R., and Bernard E. Trainor. *Cobra II: The Inside Story of the Invasion and Occupation of Iraq*. New York: Pantheon Books, 2006.
Gorman, Siobham. "CIA's Drones, Barley Secret, Receive Rare Public Nod." *Wall Street Journal*, 11 February 2014.
Gorst, Isabel. "Russian Military Budget Sapped by Corruption." *Financial Times*, 24 May 2011.
The Government of New Zealand. *Immigration and Protection Tribunal Case Number [2012] NZIPT 800151*. http://www.refworld.org/pdfid/50321fd92.pdf.
Grau, Lester, and Timothy Thomas. "Russian Lessons Learned in the Battles for Grozny." *Marine Corps Gazette*, April 2000.
Greeley, Ann T. "Psychology, Technology, and the Art of Expert Witness Persuasion in the Internet Age." *Dispute Resolution Insights* (Summer 2011).
Greenburg, Jon. "John Kerry Tells CBS Viewers that Russian Isolation Comes with Big Economic Risks." *Tampa Bay Times*, undated.
Greenspan, Alan. *The Map and the Territory*. New York: Penguin Press, 2013.
Greenwald, Glenn, and Ewen MacAskill. "Obama Orders U.S. to Draw Up Overseas Target List for Cyber-attacks." *The Guardian*, 7 June 2013.
Gridneff, Ilya. "China Sells South Sudan Arms as Its Government Talks Peace." *Bloomburg News*, 9 July 2014.
Guardian Staff Report. "Edward Snowden and the NSA Files—Timeline." *The Guardian*, 23 June 2013.
Hale, Robert F. "Promoting Efficiency in the Department of Defense: Keep Trying, Be Realistic." Center for Strategic and Budgetary Assessments, January 2002.
Hanlon, Mike. "A Computer for the World's 4 Billion Illiterate and Poor People." *GIZMAG*. http://www.gizmag.com/go/3091/.
Hardy, Roger. "The Iran-Iraq War: 25 Years On." BBC News. 22 September 2005.

Bibliography

Hardy, Seth. "IExplore RAT." *Citizen Lab Technical Brief*, August 2012. https://citizen lab.org/wp-content/uploads/2012/09/IEXPL0RE_RAT.pdf.
Harel, Amos, and Gili Cohen. "Israeli Military Seeks Massive Top-up of 2014 Defense Budget." *Haaretz*, 6 December 2013. http://www.haaretz.com/news/diplomacy-defense/.premium-1.555217.
_____. "Reports: Israeli Planes Attack Hezbollah Targets on Lebanon-Syria Border." *Haaretz*, 25 February 2014.
Harold, Scott, and Alireza Nader. "China and Iran—Economic, Political, and Military Relations." RAND Corporation, 2012.
Harris, Shane, and Mathew Aid. "Exclusive: CIA Files Prove America Helped Saddam as he Gassed Iran." *Foreign Policy*, 26 August 2013.
Haughton, Dominique, et al. "A Review of Software Packages for Data Mining." *The American Statistician* 57, no. 4 (November 2003).
Hedges, Laura. "Tele2 Russia to Deploy LTE Network in Crimea." *Capacity* magazine, 2 April 2014.
Heritage, Timothy. "Putin Dissolves State News Agency, Tightens Grip on Russia Media." Reuters, 9 December 2013.
Hern, Alex. "North Korean 'Cyberwarfare' Said to Have Cost South Korea L500M." *Guardian*, 16 October 2013.
Herszenhorn, David M., and Andrew Roth. "In East Ukraine, Protesters Seek Russian Troops." *New York Times*, 7 April 2014.
Heuer, Richard J., Jr. *Psychology of Intelligence Analysis*. Central Intelligence Agency.
Hille, Kathrin. "Russia Bars Ukraine Banks from Crimea." *Financial Times*, 21 April 2014.
_____. "Russian Bill Raises Web Freedom Fears." *Financial Times*, 5–6 July 2014.
Hinshaw, Drew, Adama Diakite, and Stacy Meichtry. "France Faces Revival of Mali Militants." *Wall Street Journal*, 19 December 2013. http://online.wsj.com/news/article_email/SB10001424052702303330204579246103662992322-lMyQjAxMTAzMDIwMDEyNDAyWj.
Hoenig, Milton. "Hezbollah and the Use of Drones as a Weapon of Terrorism." Federation of American Scientists, 5 June 2014.
Holmes, Pernilla. "The Art of Photojournalism." HowtoSpendIt (website of the *Financial Times*), 7 April 2014. http://howtospendit.ft.com/art/51443-the-art-of-photojournalism.
Hopkins, Nick. "Hackers Have Breached Top Secret MoD Systems, Cyber-security Chief Admits." *The Guardian*, 3 May 2012.
Hosenball, Mark. "Exclusive: Obama Authorizes Secret U.S. Support for Syrian Rebels." *Reuters*, 1 August 2012.
IANS. "Hackers Attack Russian Media Outlets." *Business Standard*, 18 April 2014.
Ike, Steven. "KGB Influence 'Soars Under Putin.'" BBC News, 13 December 2006.
Information Warfare Monitor and Shadowserver Foundation. "Shadows in the Cloud: Investigating Cyber Espionage 2.0." 6 April 2010.
Intelligence and Security Committee Annual Report to Parliament. July 2013.
International Committee of the Red Cross. *Methods and Means of Warfare*. http://www.icrc.org/eng/war-and-law/conduct-hostilities/methods-means-warfare/overview-methods-and-means-of-warfare.htm.
International Telecommunications Union (Geneva). "Percentage of Individuals Using the Internet 2010–2012." June 2013, retrieved 22 June 2013.
The Internet Society Studies. Global Internet User Survey 2012. http://www.internet society.org/surveyexplorer/key_findings.
ISH Janes. "Ratification of Russian Military Base Deal Provides Tajikistan with Important Security Guarantees." Janes 360, 1 October 2013.
ITAR-TASS News Agency. "Hackers Crack PrivatBank Systems for Sponsoring Military Campaign in Ukraine." 30 June 2014.

———. "Russian Lawmaker Valery Seleznyov Confirms Detention of His Son by U.S. Secret Service." 8 July 2014.
Jackson, Brian A. *Considering the Creation of a Domestic Intelligence Agency in the United States*. Santa Monica, CA: RAND Corporation, 2009.
Jacobs, Lawrence R., and Benjamin I. Page. "Who Influences U.S. Foreign Policy?" *American Political Science Review* 99, no. 1 (February 2005).
Jamjoom, Mohammed. "Source: 'Massive' Attack on Targets al Qaeda in Yemen." CNN, 20 April 2014.
Jarecki, Eugene. *Why We Fight*. Sony Pictures Classics, 2006.
Jones, Sam. "Evidence Points to Insurgent Puppetmaster." *Financial Times*, 9 August 2014.
———. "Photos and Roses for GRU's 'Spetsnaz' Casualties." *Financial Times*, 8 August 2014.
Kaiman, Jonathan. "Six Injured in Knife Attack at Chinese Train Station." *The Guardian*, 6 May 2014.
Karnitschnig, Matthew. "German Businesses Urge Halt on Sanctions Against Russia." *Wall Street Journal*, 1 May 2014.
Kates, Glenn. "Ukraine's Minister of Facebook." Radio Free Europe, 9 July 2014.
Karsh, Efraim. *The Iran-Iraq War: 1980–1988*. London: Osprey, 2002.
Keck, Zachary. "China Is a Leading Proliferator of Small Arms." *The Diplomat*, October 2013. http://thediplomat.com/2013/10/china-is-a-leading-proliferator-of-small-arms/.
Kim, Cheryl. "Think You Are a Thought Leader? You Are Probably Wrong ... but Here Are Three Ways to Become One." *Financial Post*, 7 March 2014.
Kinder, Tabatha. "Israel Uses Image of London Missile Strike to Appeal for Gaza Support." *International Business Times*, 21 July 2014.
King, Gary, Jennifer Pan, and Margaret Roberts. "How Censorship in China Allows Government Criticism but Silences Collective Expression." *American Political Science Review* 107, no. 2 (May): 1–18.
Kramer, Andrew, and Michael Gordon. "Russia Sent Tanks to Separatists in Ukraine, U.S. Says." *New York Times*, 13 June 2014.
Krebs, Brian. "Hackers Plundered Israeli Defense Firms That Built 'Iron Dome' Missile Defense System." KrebsonSecurity, 14 July 2014. http://krebsonsecurity.com/2014/07/hackers-plundered-israeli-defense-firms-that-built-iron-dome-missile-defense-system/.
———. "Russian Business Network, Down but Not Out." *Washington Post*, 7 November 2007.
———. "Shadowy Russian Firm Seen as Conduit for Cybercrime." *Washington Post*, 13 October 2013.
Kruger, Lennard G. "Internet Governance and the Domain Name System: Issues for Congress." Congressional Research Service, 10 June 2014.
Lee, David. "Russia and Ukraine in Cyber 'Stand-off.'" *BBC News Technology Report*, 5 March 2014.
Lee, Michael J., et al. "Network Discovery with Multi-intelligence Sources." *Lincoln Laboratory Journal* 20, no. 1 (2013).
Lenin, Vladimir. *War and Revolution*. Pravda, 23 April 1917. As published in *Lenin Collected Works*, Vol. 24. Moscow: Progress, 1964.
Levitt, Matthew, and Michael Jacobson. "The Money Trail: Finding, Following and Freezing Terrorist Financing." The Washington Institute, November 2008.
Lewis, James, and Stewart Baker. "The Economic Impact of Cyber Crime and Cyber Espionage." Center for International and Strategic Studies, July 2013.
Libya Television. "Kidal: Lastest Tuareg Mercenaries (MIA, MNLA) Preparing for Ultimate Battle." Al Jazeera English, 13 May 2013.
"Literacy by Country." World by Map. http://world.bymap.org/LiteracyRates.html.

Lo Ping Cheng Ming. "Secrets About CPC Spies—Tens of Thousands of Them Scattered Over 170-Odd Cities Worldwide." No. 231, 1 January 1997, 6–9.
Lockheed Martin Annual Report. 2012.
Londofio, Ernesto. "As U.S. Withdraws from Afghanistan, Poppy Trade It Spent Billions Fighting Still Flourishes." *Washington Post*, 3 November 2013.
MacAskill, Ewen. "Does U.S. Evidence Prove Russian Special Forces Are in Eastern Ukraine?" *The Guardian*, 22 April 2014.
_____, Julian Borger, Nick Hopkins, et al. "GCHQ Taps Fibre-optic Cables for Secret Access to World's Communications." *The Guardian*, 21 June 2013.
MacKenzie, Jean. "Who Is Funding the Afghan Taliban? You Don't Want to Know." Reuters, 13 August 2009.
MacKinnon, Leslie. "The Mystery of Irwin Cotler's Poisoning in Russia." *CBC*, 1 April 2014.
MacLeod, Caleb. "No Guns, Just Knives: Chilling Details of 'China's 9/11.'" *USA Today*, 30 March 2014.
Mao Tse-tung. "Be Concerned with the Well-Being of the Masses, Pay Attention to Methods of Work" (January 27, 1934), Selected Works, Vol. I, p. 147.
Marcus, Jonathan. "Ukraine: The Military Balance of Power." BBC News, 3 March 2014.
Marcus, Lauri Lowenthan. "'Jews Must Register' Flyer in Ukraine an Echo of Babi Yar." *Jewish Press*, 18 April 2014.
Maretinez, Michael, Talal Abu Rahma, and Kareem Khadder. "Israel Fires on 29 'Terror Sites' After Rockets from Gaza Hit Populated Areas." CNN, 12 March 2014.
Marinelog. "U.S. Sanctions Putin's Shipbuilder and Its Bank." Accessed 30 July 2014. http://www.marinelog.com/index.php?option=com_k2&view=item&id=7072:us-sanctions-putins-shipbuilder-and-its-bank&Itemid=231.
Marquand, Robert, and Ben Arnoldy. "China's Hacking Skills in Spotlight." *Christian Science Monitor*, 16 September 2007.
Marquis-Boire, Morgan, Jakub Dalek, et al. "Planet Blue Coat: Mapping Global Censorship and Surveillance Tools." *The Citizen Lab*, 15 January 2013.
Marson, James. "Russian Muscle Turns Tide, Leaving Ukraine Few Options." *Wall Street Journal*, 3 September 2014.
Martina, Michael. "Chinese State Media Says Five Suicide Bombers Carried Out Xinjian Attack." Reuters, 23 May 2014.
Mauldin, William, and Paul Sonne. "U.S. Escalates Sanctions Against Russia Over Ukraine Crisis." *Wall Street Journal*, 16 July 2014.
McDonald, Geoff, et al. "Stuxnet 0.5: The Missing Link." Symantec Security Response, 23 January 2014.
McManus, Doyle. "U.S. to Press Ukrainian on Nuclear Issue." *Los Angles Times*, 6 May 1992.
Mello, John P., Jr. "Pentagon: Yep, We Got Hacked." *Tech News World*, 26 August 2010.
Miller, Greg. "Lawmakers Seek to Stymie Plan to Shift Control of Drone Campaign from CIA to Pentagon." *Washington Post*, 15 January 2014.
_____. "Video Shows Brazen Outdoor Meeting of al-Qaeda Fighters." *Washington Post*, 15 April 2014.
Mitnick, Joshua. "Israelis Frustrated with Outcome of Gaza Conflict." *Wall Street Journal*, updated 29 August 2014.
Molander, Roger C., Andrew S. Riddile, and Peter A. Wilson. *Strategic Information Warfare: A New Face of War*. Santa Monica, CA: RAND, 1996.
Moor, Kieth. "Russian Crime Gang's $570M Cyber Theft Bid as Nation Loses $4.6B a Year to Computer Crime." *The Herald Sun*, 23 July 2014.
Morozov, Evgeny, and Joanne J. Myers. "The Net Delusion: The Dark Side of Internet Freedom." Carnegie Council, 25 January 2011.
Morris, Jessica. "Iran-Saudi Relations: A New Cold War Heating Up?" CNBC News, 28 January 2014.

Mortensen, Annie. "A Journalist Goes Missing Nearly Every Day in Ukraine." *Independent Voices*, 27 May 2014.
Murphy, Dan. "U.S. Nabs Alleged Russian Hacker—and Kremlin Cries Foul." *Christian Science Monitor*, 9 July 2014.
Nakashima, Ellen. "Pentagon Is Debating Cyber-Attacks." *Washington Post*, 6 November 2010.
———. "Report: Web Monitoring Devices Made by U.S. Firm Blue Coat Detected in Iran, Sudan." *Washington Post*, 8 July 2013.
———. "U.S. Said to Be Target of Massive Cyber-Espionage Campaign." *Washington Post*, 10 February 2013.
National Commission on Terrorist Attacks upon the United States. *The 9/11 Commission Report: Final Report of the National Commission on Terrorist Attacks Upon the United States*. New York: Norton, 2004.
National Security Agency. Cryptologic Almanac 50th Anniversary Series, Betrayers of Trust, undated.
———. Pearl Harbor Review, JN-25. http://www.nsa.gov/about/cryptologic_heritage/center_crypt_history/pearl_harbor_review/jn25.shtml.
"The 9/11 Commission Report: Identifying and Preventing Terrorist Financing." 23 August 2004.
O'Bagy, Elizabeth. *The Free Syrian Army*. Washington, DC: Institute for the Study of War, 2013. http://www.understandingwar.org/sites/default/files/The-Free-Syrian-Army-24MAR.pdf.
O'Shea, Dermot. "Real-World Drive Tests Declare a Verdict on GPS/GLONASS." *Electronic Design*, 31 May 2013.
Office of Foreign Asset Control. Sectoral Sanctions Identifications List. 29 July 2014.
Office of the Secretary of Defense. "DOD Website Defense Manpower Report." https://www.dmdc.osd.mil/appj/dwp/reports.do?category=reports&subCat=milActDutReg.
Oliver, Christian. "Transnistria Feels Squeezed by Clash Between East and West." *Financial Times*, 5/6 April 2014.
Openet Initiative. "China's Green Dam: The Implications of Government Control Encroaching on the Home PC." https://opennet.net/chinas-green-dam-the-implications-government-control-encroaching-home-pc.
Osen LLC, Counter-Terrorism. *Linde v. Arab Bank*. http://www.osenlaw.com/case/arab-bank-case.
Page, Jeremy, "Suspect Shown on Chinese TV Confessing to Ax Attack." *Wall Street Journal*, 22 June 2014.
———, and Ned Levin. "Web Preaches Jihad to China's Muslim Uighurs." *Wall Street Journal*, 24 June 2014.
Page, Jeremy, Brian Spegele, and James Areddy. "China Puts Ex-Security Chief Zhou YongKang Under Investigation." *Wall Street Journal*, 29 July 2014.
Palazzolo, Joe. "Terror Victims to Press Claims Against Arab Bank." *Wall Street Journal*, 8 August 2014.
Parfitt, Tom. "Russian Media Ignore WikiLeaks Revelations." *The Guardian*, 2 December 2010.
Perlez, Jane. "China and Russia, in a Display of Unity, Hold Naval Exercises." *New York Times*, 10 July 2013.
———. "China Shows Interest in Afghan Security, Fearing Taliban Would Help Separatists." *New York Times*, 8 June 2012.
Perlroth, Nicole. "Hackers in China Attacked the Times for the Last 4 Months." *New York Times*, 30 January 2013.
———. "In Cyberattack on Saudi Firm, U.S. Sees Iran Firing Back." *New York Times*, 23 October 2012.

_____. "Washington Post Joins List of News Media Hacked by the Chinese." *Washington Post*, 1 February 2013.

_____, and Quentin Hardy. "Bank Hacking Was the Work of Iranians, Officials Say." *New York Times*, 8 January 2013.

Pew Forum on Religious and Public Life. December 2012. http://www.pewforum.org/2012/12/18/global-religious-landscape-exec/.

Pew Research Center. *Far More Continue to View Russia as "a Serious Problem" Than as an "Adversary."* Survey, 28 July 2014. http://www.people-press.org/2014/07/28/far-more-continue-to-view-russia-as-a-serious-problem-than-as-an-adversary/.

Pew Research Global Attitude Project. *Global Opinion of Russia Mixed*, 3 September 2013. http://www.pewglobal.org/2013/09/03/global-opinion-of-russia-mixed/.

Pifer, Steven. "Ukraine's Perpetual East-West Balancing Act." Brookings Institute, 28 February 2014.

Pomerantsev, Peter. "How Putin Is Reinventing Warfare." *Foreign Policy*, 5 May 2014.

Prados, John. "The Navy's Biggest Betrayal." *Naval History Magazine* 24, no. 3 (June 2010).

Prados, John, and Svetlana Savranskaya, eds. *Volume II: Afghanistan: Lessons from the Last War*. National Security Archive (2001).

Privacy and Civil Liberties Oversight Board. 23 January 2014. http://www.fas.org/irp/offdocs/pclob-215.pdf.

Promotional piece. "Analysis: DREAM Project Expected to Enhance Europe-Asia Connectivity." *Capacity* magazine, 2013.

Public Broadcasting Service. *The Iranian Hostage Crisis*, n.d. http://www.pbs.org/wgbh/americanexperience/features/general-article/carter-hostage-crisis/.

Radio Dabanga (Sudan). "Sudan's Military Industry Expanding: Small Arms Survey." 6 July 2014. http://www.sipri.org/research/armaments/production/Top100.

Raviv, Dan, and Yossi Melman. *Every Spy a Prince*. Boston: Houghton Mifflin, 1990.

Rawnsley, Adam. "Espionage? Moi?" *Foreign Policy*, 1 July 2013.

Raytheon Annual Report. 2012.

Reilly, Jill. "The Guardian Allowed GCHQ Spies into Its Offices to Oversee Destruction of Leaked NSA Data After 'Senior Government Figure Backed by Prime Minister' Threatened Legal Action." *The Daily Mail*, 20 August 2013.

Reitman, Janet. "The Rise and Fall of Jeremy Hammond: Enemy of the State." *Rolling Stone*, 7 December 2012.

Reuters. "Update 3—Russia Targets Space Station Project in Retaliation for U.S. Sanctions." 13 May 2014.

Reuters and Al Jazeera. "Yemen Drones Strikes, Ambushes Kill 10." 3 March 2014.

Reuters Staff Report. "Edward Snowden Tells Der Spiegel NSA Is 'in Bed with the Germans.'" *The Guardian*, 7 July 2013.

RIA Novosti. "Corruption Up 450% in a Year in Russian Military—Prosecutors." 7 November 2013.

_____. "Two-Thirds of Russians Are Internet Users." 21 May 2013.

Richelson, Jeffrey. *Iraq and Weapons of Mass Destruction*. [Washington, D.C.]: National Security Archive, 2002. http://www2.gwu.edu/%7Ensarchiv/NSAEBB/NSAEBB80/#doc15.

Robehmed, Natalie. "U.S. Treasury Sanctions Four Russian Billionaires for Ukraine Involvement." *Forbes*, March 2014.

RT. "U.S. Seen More Risky for Industrial Espionage: German Study." 6 August 2013.

_____. Unattributed Russian article with video: "Yanukovich Sent Letter to Putin Asking for Russian Military Presence in Ukraine." 3 March 2014.

Sanger, David. "Obama Order Sped Up Wave of Cyberattacks Against Iran." *New York Times*, 1 June 2012.

———, David Barboza, and Nicole Perlroth. "Chinese Army Unit Is Seen as Tied to Hacking Against U.S." *New York Times*, 18 February 2013.

Sanger, David E., and Steven Erlanger. "Suspicion Falls on Russia as 'Snake' Cyberattacks Target Ukraine's Government." *New York Times*, 8 March 2014.

Schlesinger, Arthur M., Jr. "The Measure of Diplomacy." *Foreign Affairs*, July/August 1994. http://www.foreignaffairs.com/articles/50118/arthur-m-schlesinger-jr/the-measure-of-diplomacy.

Schneier, Bruce. Congress for Privacy and Security. http://www.youtube.com/watch?v=Skr-jIqISO0.

Schwartz, Moshe. "Twenty-Five Years of Acquisition Reform: Where Do We Go from Here?" Congressional Research Service, 29 October 2013.

Security Service of the Ukraine. "A Military Spy Detained" (in English). 14 March 2014. http://www.sbu.gov.ua/sbu/control/en/publish/article?art_id=122723&cat_id=35317.

Shadowserver Foundation Report AS40989. "RBN as RBusiness Network." 6 January 2008.

Shanker, Thom. "U.S. Arms Sales Make Up Most of Global Market." *New York Times*, 26 August 2012. http://www.nytimes.com/2012/08/27/world/middleeast/us-foreign-arms-sales-reach-66-3-billion-in-2011.html.

Sharp, Jeremy M. "U.S. Foreign Aid to Israel." Congressional Research Service, 11 April 2013. http://www.fas.org/sgp/crs/mideast/RL33222.pdf.

Shell, Elizabeth, and Matt Stiles. "Where Does U.S. Military Aid Go?" McNeil/Lehrer Productions, 30 August 2012. http://www.pbs.org/newshour/spc/multimedia/military-spending/.

Shogren, Elizabeth. "Tensions Rise Over Fate of Black Sea Fleet." *Los Angeles Times*, 4 April 1992.

Silverman, Jacob. "Loose Tweets Sink Ships." *Politico*, 28 August 2014.

Skillings, Jonathan. "Airborne Laser Hits the Off Switch." CNET, 27 February 2012.

Smith, R. Jeffrey, and Joby Warrick. "Pakistani Scientist Khan Describes Iranian Efforts to Buy Nuclear Bombs." *Washington Post*, March 14, 2010.

Sneed, Adam. "Rogers: Russia May Be Behind Snowden Leaks." *Politico*, 14 January 2014.

Soldatov, Andrei. "Vladimir Putin's Cyber Warriors." *Foreign Affairs*, 9 December 2011.

———, and Irina Borogan. "The Kremlin's Internet Surveillance Plan Goes Live Today." *Wired*, 1 November 2012.

Solomon, Jay. "Ties to Russia Arms Supplier Snarl U.S. Sanctions Effort." *Wall Street Journal*, 28 March 2014.

———. "U.S. Blacklists Companies That Supplied Hezbollah." *Wall Street Journal*, 11 July 2014.

Sonne, Paul, Philip Shishkin, and Anton Troianovsk. "Shift Toward Moscow Jolts Crimean Economy." *Wall Street Journal*, 17 March 2014.

Star, Alexander, and Bill Keller. *Open Secrets: WikiLeaks, War and American Diplomacy*. New York: Grove Press, 2011.

Stecklow, Steve. "Dubai Firm Fined $2.8 Million for Shipping Blue Coat Monitoring Gear to Syria." Reuters, 25 April 2013.

Sterling, Bruce. "Russian Business Network Sets Up Shop in China." *Wired*, 9 November 2013.

The Stockholm International Peace Research Institute. "The SIPRI Top 100 Arms-Producing and Military Services Companies in the World Excluding China, 2011." 2013.

The Stockholm International Peace Research Institute. "The SIPRI Yearbook." 2013.

Summerville, Quinton. "Pakistan Helping Afghan Taliban—NATO." BBC News, 1 February 2012.

Surveillance. Center for Strategic and International Studies, 18 April 2014.

Sutter, John D. "When the Internet Actually Helps Dictators." Cable News Network, 22 February 2011.

Swartz, Phillip. "Lawmakers Force Pentagon to Buy Tanks, Keep Ships and Planes It Doesn't Need." *Washington Times*, 9 May 2013.

Schwarzkopf, H. Norman, and Peter Petre. *It Doesn't Take a Hero: General H. Norman Schwarzkopf, the Autobiography*. New York: Bantam Books, 1992.

Sweeney, John, Jens Holsoe and Ed Vulliamy. "NATO Bombed Chinese Deliberately." *The Observer*, 16 October 1999.

Symantec anonymous employee. "Emerging Threat: Dragonfly/Energetic Bear—APT Group." Symantec Corp, 30 June 2014. http://www.symantec.com/connect/blogs/emerging-threat-dragonfly-energetic-bear-apt-group.

Synovitz, Ron. "Tracking the Terrorist Money Trail." Radio Free Europe/Radio Liberty, November 2008. http://www.rferl.org/content/Interview_Tracking_The_Terrorist_Money_Trail/1349657.html.

Szczechowski, Master Sgt. Jeff, USAF. "Security Forces Help Keep Convoys Safe in Afghanistan." American Forces Press Service, 26 May 2004.

Tanas, Alexander. "Moldova Blocks Russian Plan to Expand Presence in Rebel Enclave." Reuters, 17 November 2012.

Taubman, Philip. "Soviets List Afghan War Toll: 13,310 Dead, 35,478 Wounded." *New York Times*, 26 May 1988.

Thompson, Clive. "The Darkest Place on the Internet Isn't Just for Criminals." *Wired*, 18 October 2013.

Thomsen, Alexandra. "The Disease Daily." *Healthmap*, July 2014.

Thornburgh, Nathan. "Inside the Chinese Hack Attack." *Time*, 25 August 2005.

Tisdall, Simon. "Kurdistan Faces Long, Fraught Road to Sustainable Independence," *The Guardian*, 3 July 2014.

Torbenson, Eric. "ATC Problems on East Coast: Not Much Here." *Dallas Morning News*, 19 November 2009.

Traynor, Ian. "Russia Accused of Unleashing Cyberwar to Disable Estonia." *The Guardian*, 16 May 2007.

Troianovski, Anton. "Russia Ramps Up Information War in Europe." *Wall Street Journal*, 21 August 2014.

Ukrtelecom Press Release. "Ukrtelecom's Crimean Sub-branches Officially Report That Unknown People Have Seized Several Telecommunications Nodes in Crimea." 28 February 2014.

United Press International. "Russia Plans Increases in Spending on Defense, Nuclear Weapons." 8 October 2013. http://www.upi.com/Top_News/World-News/2013/10/08/Russia-plans-increases-in-spending-on-defense-nuclear-weapons/UPI-36261381251991/.

United Press International Business News. "DSCA Outlines Foreign Military Sales Program." 20 September 2013. http://www.upi.com/Business_News/Security-Industry/2013/09/20/DSCA-outlines-foreign-military-sales-program/UPI-41501379711418/-ixzz2uGUs4X00.

U.S. Army. "Mobilization." Undated, page 21. http://www.history.army.mil/documents/mobpam.htm.

U.S. Department of Justice, Press Release. "Manhattan U.S. Attorney Announces Guilty Plea of Jeremy Hammond for Hacking into the Stratfor Website." 28 May 2013.

U.S. Department of the Treasury. "Treasury Targets Key Al-Qa'ida Funding and Support Network Using Iran as a Critical Transit Point." 28 July 2011.

———. "U.S. Designates Five Charities Funding Hamas and Six Senior Hamas Leaders as Terrorist Entities." 22 August 2003.

Urquhart, Conal. "Summary of Ukraine Events." *The Guardian*, 1 March 2014.

Vassilieva, Maria. "Russian Elections: Hunting the 'Carousel' Voters." *BBC Russia*, 5 March 2012.
Vatis, Michael. "Cyber Attacks: Protecting America's Security Against Digital Threats." ESDP Discussion Paper ESDP-2002-04. John F. Kennedy School of Government, Harvard University, June 2002.
Villeneuve, Nart, Ned Moran, and Thoufique Haq. "Operation Molerats: Middle East Cyber Attacks Using Poison Ivy." *FireEye*, 23 August 2013.
Vodaphone. Law Enforcement Disclosure Report. 2014. http://www.vodafone.com/content/sustainabilityreport/2014/index/operating_responsibly/privacy_and_security/law_enforcement.html#eocp.
Voice of Russia. "U.S. Starts World Press Freedom Day with Even More Lies." 6 May 2014.
Walsh, Declan. "Pakistan Suspends License of Leading News Channel." *New York Times*, 7 June 2014.
Walsh, Michael. "Unmasked." Smithsonian Institution, May 2009.
Walton, Greg. "China's Golden Shield." International Center for Human Rights and Development, 2001.
Ward, Stephen. "An Iranian Threat Inside Social Media." iSight Partners, 28 May 2014.
Waters, Richard. "Attacks Lead to New Line for Telcos in U.S." *The Financial Times*, 12 October 2001.
Weaver, Courtney. "Moscow Launches Mac Attack as Beef with West Over Ukraine Intensifies." *Financial Times*, 26–27 July 2014.
Weber, Lauren. "At Work: Remote Data Wipes of Workers' Personal Devices Are Rising." *Wall Street Journal*, 9 September 2014.
Weinberger, Sharon. "How to Do Business with a Blacklisted Russian Weapons Company." *Wired*, 28 July 2008.
Weiss, Michael. Cable News Network, a Turner Broadcasting System of Time Warner. 18 July 2014.
Weitz, Richard. "Superpower Symbiosis: The Russia-China Axis." *World Affairs Journal*, November–December 2012.
WGBH Educational Affiliation. *Nixon's China Game, American Experience*. Public Broadcasting, 1999.
Whatley, Frederick W. *Reagan, National Security, and the First Amendment: Plugging Leaks by Shutting Off the Main*. Washington, DC: Cato Institute, 1984.
Wolosky, Lee S. "How to Sanction Russia." *Foreign Affairs*, Council on Foreign Relations, 19 March 2014.
Zakaria, Fareed. "Russian and Chinese Assertiveness Poses New Foreign Policy Challenges." HBO History Makers Series. Council on Foreign Relations, 21 May 2014.
Zarate, Juan C., and Thomas Sanderson. "How the Terrorists Got Rich." *New York Times*, 28 June 2014.
Zero Dark Thirty. Sony Pictures, 2013.
Zittrain, Jonathan, and Benjamin Edelman. "Empirical Analysis of Internet Filtering in China." Harvard Law School, March 2003.

INDEX

Abkhazia 15
Absentee ballots and carousel voting 19
Afghanistan war, annual cost of 119
Agent BTZ 104, 137
Al Jazeera 54, 55, 71, 108, 119, 197–201
Al Qaeda 16, 47, 56, 107–109, 117–121, 123–124
Ames, Aldrich 170–171
Anonymous 75, 165
Anti, Michael 74
Aoussa, Cheick 62, 119–120
Ashton, Kevin 135
Autocratic kleptocracy 141
Avakov, Arsen 14

Babi Yar 35
al-Badri al Sammarai, Awwad Ibrahim Ali 47
Balkanization of the Internet 4, 88, 182
Bamford, James 168–169
Banned ways and methods of warfare 14
Ben Gurion Airport 54
bin Laden, Osama 16, 19, 46
Blair, Tony 126
Bluecoat 76, 84
BNP Paribas 36–37
Boisvert, Ray 24
Bronze Soldier statue 25

Cable News Network (CNN) 4, 11, 31, 50–55
Caihou, General Xu 145
Cell phones, jamming 35
Censorship bureau 59, 71
Chalabi, Ahmed 48
Chechnya 10, 13, 16, 118, 120, 134
Chicken Kiev speech 22
China-Sudan Arms manufacturing 17, 36, 154–155
Code 200 20

Code fragments 183
Command and Control Warfare 3, 5, 104–105, 107
Committee of Soldier's Mothers 13, 32
Computer Network Attack (CNA) 133, 157–160, 166
Computer Network Exploitation (CNE) 133, 157, 160, 166, 173
Credibility 1, 11–12, 15, 20, 24, 32, 38, 51–53, 112
Crony capitalism 151
Cutler, Irwin 24
Cyber Arms Merchants 25, 58, 108, 155, 157
Cyberwar 1–7, 11–12, 14, 16, 18–21, 35, 57, 67–79; definitions 1–5

DBN 96
Defense contractors 65, 89, 91–92, 107, 137–138
Deniability 11–12, 20, 38, 173; *see also* Operational plausible deniability
Denial *see* Service attacks, denial of
Deterrence 24–25, 127, 132–133, 185
Door-knocking 54
Dragonfly/Energetic Bear 81–82, 92, 95
DREAM (Network Diverse Route for European and Asian Markets) 33
Drones (HAMAS) 53, 84

Economic espionage 58, 86–87, 99, 101, 103
Economic Warfare 3, 5, 33, 35–36, 38, 67, 92–93, 96, 141, 162, 175
Eisenhower, Dwight D. 147–148
Electronic Warfare 3, 5, 105, 150
Espionage *see* Economic espionage; French industrial espionage
Estonia 17, 25, 55, 94, 176
Europe-Persian Gateway 30

Facebook 10, 15, 19, 27–28, 69, 73–76, 79, 83, 88, 90, 139
Farr, Charles 161
The Federal Security Service of the Russian Federation (FSB) 24, 27, 34, 38, 77, 102, 121, 142–145, 168
Federally Funded Research and Development Contractors 156
First targets 10
Five-eyes countries 55, 156
Foreign Military Sales (FMS) 150, 165
French industrial espionage 97, 101
Friends of Putin 40–41

Gansler, Jackques 148
Gates, Robert M.: on Russia 178
Gazprom 35
Geneva Convention 46–47
Geo News 31
Georgia 15, 17
Glonass 17
Golden Shield 77–78
Google 48, 62, 77–79, 88, 90, 98, 102, 108, 119, 139–140, 181–182
Great Cannon 77–78
Green Dam 78, 80
Green Dragon Crew 14
Green Revolution 74

Hacker Defense 93–94
Hacking Iron Dome manufacturers 156
Hammond, Jeremy 165
Hanssen, Robert 171
HBGary Federal 165
Heartbleed 136
Hezbollah 56, 109, 126–127, 129
Highway of Death 45
Humphrey, Hubert: Russian involvement in campaign 26

Infrastructure disruptions 10, 30, 39, 57, 61, 94–95, 100, 130–132, 184–185, 190
"Inherently government functions" 158–159
Interception *see* Lawful interception
Internet as a CIA project 182
Internet Containment 179
Internet Corporation for Assigned Names and Numbers (ICANN) 136, 182
Internet of Things 135–136
Internet Police 83, 137
Inter-Service Intelligence (ISI) 31, 121, 123
Iranian YouTube video 127
Iraqi General Samarrai 43
Iron Dome 53, 67, 156
ISIS/ISIL 2, 47–49, 56, 74, 109, 117–118, 124, 133, 151, 175, 194

Jacobson, Michael 124
Jamming *see* Cell phones, jamming
Al Jazeera 54, 55, 71, 108, 119, 197–201
Johnson, Stanley 106
JustPaste.It 74

KAL 007 51
Khan, Abdul Qadeer 128
Khodorkovsky, Mikhail B. 103
Khomeini, Ruhollah 41
Kolomoyski, Ihor 14
Koval, George 170–172
Kravchuk, Leonid M. 22–23
Krysha (roof) 143

Lawful interception 70–71
Lenin, Vladimir 48, 125
Literacy factor 111–112
Lloyd, Odin 80

Mandiant, Inc. 98–100, 135
Manhattan Project 170
Manning, Chelsea 186
Martin, William, and Bernon Mitchell 170
Maskhadov, Aslan 16
McCaul, Michael 168
McLuhan, Marshall 114
Media, control of 2, 4–5, 19, 21, 27, 50, 57, 59, 67–69, 73–75, 77, 80, 83, 87–88, 90, 100, 112–115, 176, 179, 185, 190–191
Medvedev, Dmitry 30, 36, 142, 146
MI-17 helicopters 18
Moldova 15
Mondale, Vice President Walter F. 41
Moonlight Maze 100
Morozov, Evgeny 73–74
Morsi, Mohammad 112, 114
Mosul 47–48

NarusInsight 84
National Defense budget 64–65
National Intelligence Estimate (Economic Espionage) 97
NATO–Russia Council 56
Neo-Nazi Theme 5, 10
Network mapping 85, 138, 176, 182
Newscaster 89
Non-classified Internet Protocol Router Network (NIPRNET) 100, 137–138, 146
North Atlantic Treaty Organization (NATO) 12, 14, 17, 23, 25, 34, 50, 55–56, 63–65, 151
Nuclear proliferation 56, 63

Official Secrets Act 186
Operation Ababil 57, 90

Operational plausible deniability 11–12, 20, 173
Opinions, bell-shaped curve of 2, 32
Osen law firm 122
Ossetia 15–16

Palestinians 53–54, 124, 152
Patriot Act 70
Perception Management 177–178
Poison Ivy 167
Presidential Decision Directive 63, 130
Presidential Policy Directive 20, 28, 71, 157
PrivatBank 14
Propaganda 5, 33, 183
Psychological Warfare 3, 5, 110–115

Al Qaeda 16, 47, 56, 107–109, 117–121, 123–124

RAND Secretary of Defense Study on the New Face of War 176–177, 179, 181, 185
Reasonable doubt 20, 94
Remote Access Tools (RATS) 81–82, 183
Revolutionary defencism 125
Rogers, Mike 168
Roskomnadzor 77
Rosoboronexport 17–18
Rospotrebnadzor 17
Rostelecom 30
Russia Segodnya 73
Russian Business Network (RBN) 85–87
Russian Central Bank 38
Russian Nationalists 15, 18, 20–21, 32, 50–52
Russian Organized Crime Group (OCG) 143

SA-11(anti-aircraft system) 50–53
Saman, Moises 111
Sanctions 16–18, 36–38, 51, 94, 101, 175
SANS 166
Saudi Aramco 89
Saudi Committee for the Support of the Intifada al Quds 121–122
Schlesinger Arthur, Jr. 66
Schneier, Bruce 79–80
Schwarzkopf, General Norman 43, 185
Section 215 of the USA PATRIOT Act 70
Seleznyov, Roman 115–116
Service attacks, denial of 72, 90, 94, 98
SHADOW 82
Shaya, Uzi 121
Sheep dipping 52
Shpigun, General Gennady Nikolaevich 16
Silk Road website 60–61, 191
Snake/Agent.btz 104, 137
Snowden, Edward 30, 60, 68, 87, 103, 115, 140, 142, 158–159, 167–171

Solar Sunrise 100
Special forces 12–13, 19–20, 29, 35, 42, 44–45, 63, 115, 148, 158
Specially Designated Global Terrorists 122
Spetsnaz 13
Spying 25–26, 59, 67, 70, 87, 103, 117, 168–170
Stars Group Holding 126
Stuxnet 61, 63, 87, 95, 127, 133
System for Operative Investigative Activities (SORM) 77, 80

T-64 tank 14
Telecomix 76
Telvent 135
Terrorist financing 121–122, 124, 191, 196
Thought leader 19, 25–27, 29, 33, 67, 73, 102, 127, 142
Threat assessment 56–57, 60
Titan Rain 100
Tomahawk missile 44, 153
Transnational crime 56, 60
Twitter 4, 28, 73–74, 76–77, 83, 87–88, 90, 113
Tymoshenko, Yulia 23, 34

Uighurs 120–121
Ukrtelecom 26
Uppsala University Conflict database 40, 60
Ussuri River 64

Verizon Communications vs the Federal Communications Commission 181
Vice Media 139
Vodaphone disclosure 69–71, 134
Von Clausewitz, Carl 2, 5, 40–41, 65–66, 68, 145, 147
Voter fraud and carrousel voting 23

Walker, John 170–171
Weapons of Mass Destruction in Iraq 43
Weiss, Michael 38
Weitz, Richard 63–64
Wen, Jiabao 89
Will of the people 2–6, 18
Wolosky, Lee 38
Work factor analysis 26
al-Wuhayshi, Nasir 107

Xi, Jinping 145
Xinjiang 72, 121

Yanukovych, Viktor 10–11, 25, 33–35, 44
Yeltsin, Boris 16, 22–23
Yushchenko, Viktor 23, 194

Zhou, Yongkang 146

www.ingramcontent.com/pod-product-compliance
Ingram Content Group UK Ltd.
Pitfield, Milton Keynes, MK11 3LW, UK
UKHW041953140426
5217IPUK00015B/784